Philosophy of Science:
Key Thinkers

Continuum *Key Thinkers*

The *Key Thinkers* series is aimed at undergraduate students and offers clear, concise and accessible edited guides to the key thinkers in each of the central topics in philosophy. Each book offers a comprehensive overview of the major thinkers who have contributed to the historical development of a key area of philosophy, providing a survey of their major works and the evolution of the central ideas in that area.

***Key Thinkers* in Philosophy available now from Continuum*:**

Aesthetics, edited by Alessandro Giovannelli
Epistemology, edited by Stephen Hetherington
Ethics, edited by Tom Angier
Philosophy of Religion, edited by Jeffrey J. Jordan

Philosophy of Science: The Key Thinkers

Edited by
James Robert Brown

continuum

Continuum International Publishing Group

The Tower Building   80 Maiden Lane
11 York Road   Suite 704
London SE1 7NX   New York NY 10038

www.continuumbooks.com

© James Robert Brown and Contributors 2012

British Library Cataloguing-in-Publication Data
A catalogue record for this book is available from the British Library.

ISBN: HB: 978-1-4411-2881-2
      PB: 978-1-4411-4200-9

Library of Congress Cataloging-in-Publication Data
Philosophy of science : the key thinkers / [compiled by] James Robert Brown.
    p. cm. – (Key thinkers)
Includes index.
ISBN 978-1-4411-2881-2 (hardcover) – ISBN 978-1-4411-4200-9 (pbk.) –
ISBN 978-1-4411-8665-2 (ebook (pdf)) – ISBN 978-1-4411-5254-1 (ebook (epub))
1. Science–Philosophy–History. 2. Philosophers–Biography. I. Brown, James Robert.
  Q175.P51295 2012
  509.2′2–dc23

                                                        2011031385

Typeset by Newgen Imaging Systems Pvt Ltd, Chennai, India
Printed and bound in India

Contents

Acknowledgements

My first thanks go to the people at Continuum for inviting me to edit this book and for seeing it through production, especially Sarah Campbell and Tom Crick. They have been a great pleasure to work with. Perhaps my biggest thanks goes to the contributing authors – without them, nothing. They have risen to the occasion magnificently. Finally, thanks to Kevin Kuhl for doing the index. I am very grateful to all.

Notes on Contributors

Arun Bala is by training a physicist and a philosopher. He taught history and philosophy of science at the National University of Singapore and held positions as Visiting Professor in the Department of Philosophy and as adjunct professor at Trinity College, both in the University of Toronto. He is presently Visiting Senior Fellow with the Institute of Southeast Asian Studies in Singapore. He is the author of *The Dialogue of Civilizations in the Birth of Modern Science* (Palgrave Macmillan, 2008) and editor of the forthcoming book, Asia, Europe and the Emergence of Modern Science: Knowledge Crossing Boundaries (Palgrave Macmillan, 2012).

James Robert Brown is a professor of philosophy at the University of Toronto. His interests include a wide range of topics in the philosophy of science and mathematics: thought experiments, foundational issues in mathematics and physics, visual reasoning and, issues involving science and society, such as the role of commercialization in medical research. His books include *The Rational and the Social* (Routledge, 1989), *The Laboratory of the Mind: Thought Experiments in the Natural Science* (Routledge, 1991/2010), *Smoke and Mirrors: How Science Reflects Reality* (Routledge, 1994), *Philosophy of Mathematics: An Introduction to the World of Proofs and Pictures* (Routledge, 1999/2008), *Who Rules in Science: A Guide to the Wars* (Harvard, 2001), *Platonism, Naturalism and Mathematical Knowledge* (Routledge, 2011) and others forthcoming.

Martin Carrier is professor of philosophy at Bielefeld University and part of the Institute of Science and Technology Studies (IWT). His chief area of work is the philosophy of science, in particular, historical changes in science and scientific method, theory-ladenness and empirical testability, inter-theoretic relations and reductionism and, presently, methodological issues of application-oriented research. He addresses methodological changes imposed on science by the pressure of practice. Carrier is a member of the German Academy of Science Leopoldina, the Mainz Academy of Sciences, Humanities and Literature and the Academia Europaea. He was a-warded the Leibniz Prize of the German Research Association (DFG) for 2008.

Martin Curd teaches philosophy of science at Purdue University. His books include *Argument and Analysis*, *Philosophy of Science: The Central Issues* (2nd edition edited with Jan Cover and Chris Pincock) and *The Routledge Companion to Philosophy of Science* (edited with Stathis Psillos). One of his few regrets in life is that he never took a class with Carl Hempel.

Steve Fuller holds the Auguste Comte Chair in Social Epistemology at the Department of Sociology, University of Warwick. Originally trained in history and philosophy of science, he is best known for his work in the field of social epistemology (also the name of a quarterly journal he founded in 1987 and the first of his 18 books). His most recent books are *The Sociology of Intellectual Life: The Career of the Mind in and Around the Academy* (Sage, 2009), *Science: The Art of Living* (Acumen and McGill–Queens University Press, 2010) and *Humanity 2.0: The Past, Present and Future of What It Means to Be Human* (Palgrave Macmillan, 2011). He is currently working on a history of epistemology.

Janet Kourany is a fellow of the John J. Reilly Center for Science, Technology, and Values at the University of Notre Dame as well as an associate professor in Notre Dame's Philosophy Department, with special interests in issues concerned with science and values and feminist philosophy of science. Her most recent work includes *Philosophy of Science after Feminism* (Oxford University Press, 2010) and *The Challenge of the Social and the Pressure of Practice: Science and Values Revisited*, edited with Martin Carrier and Don Howard (University of Pittsburgh Press, 2008).

Martin Kusch holds a chair in theory of science and epistemology in the Department of Philosophy at the University of Vienna. Before moving to Vienna in 2009, he was professor of philosophy and sociology of science in the Department of History and Philosophy of Science at Cambridge University. His main books are *Psychologism* (Routledge, 1995), *Psychological Knowledge* (Routledge, 1999), *The Shape of Action* (with Harry Collins; MIT Press, 1998), *Knowledge by Agreement* (Oxford, 2002) and *A Sceptical Guide to Meaning and Rules* (Acumen and McGill, 2006). He is currently writing a book on Wittgenstein's challenge to contemporary epistemology.

Stathis Psillos is professor of philosophy of science and metaphysics at the University of Athens, Greece. He is the author of *Knowing the Structure of Nature* (Palgrave, 2009), *Philosophy of Science A–Z* (Edinburgh University Press, 2007), *Causation and Explanation* (McGill-Queens U.P., 2002) and *Scientific Realism: How Science Tracks Truth* (Routledge, 1999). He is also the co-editor (with Martin Curd) of *The Routledge Companion to Philosophy of Science* (Routledge, 2008). He has published more than 70 papers in learned journals and books, on scientific realism, causation, explanation and the history of philosophy of science. He has served as the president of the European Philosophy of Science Association (2007–9) and is currently the co-editor of *Metascience*.

William Seager is professor of philosophy at the University of Toronto Scarborough, where he has taught for the over 25 years. His main research interests are in the field of the philosophy of mind, especially with regard to consciousness and its relation to the scientific image of the world. He has the rather odd idea that the issue of scientific realism might be more relevant to the problem of consciousness than it appears at first glance. Publications include 'Theories of Consciousness', 'Credibility, Confirmation and Explanation' and 'A Note on the Quantum Eraser'.

Laura J. Snyder is associate professor of philosophy at St. John's University, New York City. She served as president of the International Society for the History of Philosophy of Science in 2009 and 2010 and is a co-editor of the society's journal. She is the author of numerous articles, as well as two books: *Reforming Philosophy: A Victorian Debate on Science and Society* and *The Philosophical Breakfast Club: Four Remarkable Friends Who Transformed Science and Changed the World*. Her work focuses on the ways in which discussions of scientific method are embedded in the cultural, political, intellectual and scientific contexts in which they occur.

Friedrich Stadler is professor of history and philosophy of science at the University of Vienna, and is the founder and director of the Institut Wiener Kreis/Vienna Circle Institute. He has been a visiting professor at the universities of Minnesota, Berlin and Helsinki (Collegium for Advanced Studies) and president of the European Philosophy of Science Association (EPSA). He is the author of two books, the editor of three book series and the co-editor of thirty-five books in the field of history and philosophy of science, modern intellectual history (emigration/exile studies). Book: *The Vienna Circle* (1997/2001); series editor: Ernst Mach-Studienausgabe (2008ff.); Moritz Schlick. Kritische Gesamtausgabe (2006ff.). His research projects involve the history of philosophy of science.

Torsten Wilholt is professor of philosophy and history of the natural sciences at Leibniz Universität Hannover. He studied Philosophy, Mathematics and History of Science at Göttingen and Berlin (M.A. 1998, Humboldt University) and obtained his Ph.D. in Philosophy at Bielefeld in 2002 with a dissertation on the philosophy of mathematics. In 2010, he completed his Habilitation with a thesis on the freedom of science. His publications include research papers in Philosophy of Science, Studies in History and Philosophy of Modern Physics and Studies in History and Philosophy of Science, as well as a book on the applicability of mathematics (Zahl und Wirklichkeit, Paderborn: Mentis 2004). Recent research interests include the effects that an application-oriented research agenda has on research quality and objectivity, and epistemological perspectives on the freedom of scientific research.

Preface

If you have picked up this book, you don't need to be told that science is hugely important in our lives. Its achievements have delighted and disturbed us – often at the same time. Who is not dazzled by relativity, quantum mechanics, big bang cosmology, and Darwinian evolution? The rivals of science – tradition and religious authority – can only sulk in silence or lash out in resentment at the passing parade. But not all is sweetness and light. We rightly fear many of the technological accomplishments that are linked to science. Will there be a nuclear war, perhaps an accidental one? Are genetically modified foods safe? Will the neurosciences find ways of reading our minds so that all privacy of thought is lost? These are legitimate worries, but they are not the concern of this book.

Our focus is simply this: Science is a success – why? We all know that science has something to do with reason and observation. But what exactly? This has been the occupation of philosophers, especially philosophers of science, since the beginning. This is a book about some of the key thinkers in the philosophy of science over the past couple of centuries. We could have started earlier, but it seems best to take up the subject from the time when some philosophers began to think of themselves not just as philosophers but as philosophers of science. The chapters are written by leading contemporary philosophers of science, who aim to explain, criticize, and honour the giants – the key thinkers.

The intended audience includes beginners and experts alike – philosophers, scientists and members of the broad public who care about how science works. An introductory chapter sets the stage. After that, readers can proceed in chronological order or jump around to whatever chapter strikes their fancy.

INTRODUCTION

James Robert Brown

Who were the great philosophers of science? The answer is easy: the great philosophers. Some qualification is needed, but not much. Political philosophers, for instance, wouldn't be included (though they might be as philosophers of the social sciences) and lots of great philosophers of science have been working scientists. Nevertheless, with this characterization in mind, let's take a look back.

The Pre-Socratics were clearly early scientists; they conjectured how the world works and were concerned with methodological issues involving explanation and evidence. Plato was deeply influenced by mathematics and in turn he influenced its development. Aristotle was almost certainly the greatest biologist before Darwin. Plato and Aristotle were both very influential in the methodology of science. In his dialogue *Timaeus*, Plato took the view that the heavens are inaccessible to us. We have no way of checking to see if what we say about the sun, moon, planets, and stars is really true; all we can do is 'save the phenomena'. That is, we should not aim for the truth, but only for an account of the heavens that correctly systematizes and predicts what we can actually see. Whether we say 'The earth and Mars go around the sun' or 'The sun and Mars go around the earth' is unimportant, since we could never know. What does matter is that the point of light we call Mars must be at the location and time specified by our theory. This is the phenomenon that must be saved.

Such a view – known as instrumentalism – has been popular with philosophers and scientists, especially astronomers, since the time of Plato. It stands in opposition to realism, the view that science aims for the truth and is more or less successful in getting it (at least approximately).

It is an issue that has constantly been with us. Copernicus was a realist; when he claimed that the earth and planets revolved around the sun, he meant that to be true. He died as his great work, *On the Revolutions of the Heavenly Spheres*, went to press and so the business of bringing it to the public was left in the hands of Rheticus, his student. Andreas Osiander wrote the preface anonymously. He was, among other things, sensitive to the long tradition of instrumentalism and, not wanting to ruffle church feathers, denied that the new theory was anything more than a useful way to calculate things; it was not intended to be literally true, just a good predictor. The philosophical battle was fought again by Galileo, who was a staunch realist. The church allowed him to claim that Copernicanism was a better instrument, that is, a better predictor of what we could observe, but it insisted that Galileo be forbidden to say it is true. Galileo's final years were spent in house arrest for violating this.

Plato's most famous student was Aristotle, who disagreed with many of Plato's most important doctrines. But they were on the same page for some other issues. Plato would have been happy to endorse Aristotle's famous account of causation and explanation. There are, according to Aristotle, four different types of cause and four corresponding types of explanation. A *material* cause is the stuff something is made of. A statue, to take Aristotle's example, is made of marble; it is the material cause of the statue. A *formal* cause is the idea, pattern or blueprint that the sculptor follows in making the statue. The *efficient* cause is the chiselling away by the sculptor to shape the marble into the form. And the *final* cause is the aim or goal of the statue, which might be to produce a beautiful object that people could admire or, better yet, to make something people would pay a lot to own.

The Scientific Revolution, the time when Galileo, Descartes, Boyle, Newton, Leibniz and many others were at work, brought about many changes. One of the biggest was the overthrow, not just of Aristotle's actual physics and biology, but of his philosophical outlook. This was the time when a mechanical outlook came to prevail. Gone were formal and teleological causes. Material ones remained but were not much discussed. Explanations took the form of citing efficient causes. Teleology, or final causation, remained, but only until scientists could find a way to jettison it. That came with Darwin, whose theory of evolution made no appeal to species' improving or having any sort of aim or goal. This hugely important philosophical development may be underappreciated. Current champions of Intelligent Design, who try to account for various phenomena by positing a purposive creator (God), are not just denying the details of Darwin and proposing a rival theory in its place. They are trying to reinstate final causes, a hugely regressive move, with no apparent benefit to compensate.

Discarded ideas might, however, come back into use. Intelligent Design is going nowhere, but the return of something like formal cause and formal explanation might turn out to be more fruitful. Most physical explanations make use of a causal process that gave rise to the phenomenon being explained. And that is almost invariably an efficient cause. But there are a few cases in contemporary physics where something like a formal cause or formal explanation is at work. These are explanations that appeal to some sort of general structure or pattern. Explanations in high-energy physics in terms of symmetry are plausible examples.

Aristotle was the most influential philosopher of science from his own time until the seventeenth century. His influence on medieval science and philosophy was unparalleled. Occam, Buridan and others in the middle ages concerned themselves with numerous topics that obviously belong to the philosophy of science. Buridan in particular dreamed up several brilliant thought experiments, some anticipating Galileo, that undermined Aristotelian physics and paved the way for the Scientific Revolution.

Philosophy and science have perhaps never been so intimately connected as they were during the seventeenth century, the period now called the Scientific Revolution. Galileo wrote extensively about the primary/secondary quality distinction, arguing that colours and smells and so on are a kind of illusion. Reality, he claimed, is mathematical, not sensuous. He thought deeply about the nature of scientific evidence, sometimes defending thought experiments in place of real experiments. Newton, who had strong methodological views, said, 'I do not frame hypotheses', and proposed philosophical rules. He also famously argued for absolute space and time, the stage on which material objects exist and events happen. Descartes and Leibniz are, of course, well-know philosophers. It should not be forgotten that they were also two of the greatest mathematicians of all time. Descartes created analytic geometry and Leibniz invented the calculus (at the same time Newton did). They also played a huge role in the development of physics in their time.

Locke, who thought of himself as an 'under-labourer', helped Newton by clearing up the odd conceptual confusion here and there. Berkeley is perhaps most famous for his denial of matter, but his best and most important work was as a critic of Newton. All of Mach's famous criticisms of Newton's absolute space and time were first stated by Berkeley. Also a very perceptive critic of Newton's calculus, he ridiculed the idea of an infinitesimal as the 'ghost of a departed quantity'. Hume's influence on thinking about cause and inductive inference in science, as elsewhere, goes without saying. Even Kant was a philosopher of science. According to one biographer, the vast majority of books in his personal library were on mathematics and science. Hume may have awoken him from his dogmatic slumbers, but it was mainly the hope of understanding the success of Newtonian science that seems to have motivated Kant.

In spite of this illustrious history, we think of philosophy of science as of rather recent origin. Any point we pick to begin will be arbitrary, but a fairly natural stating place is the nineteenth century, when William Whewell (pronounced *hew-el*) and John Stuart Mill clashed. Instead of talking about knowledge or evidence in general, the focus of interest became scientific knowledge, evidence in the sciences and so on. Perhaps the break was inevitable, as science was increasingly seen as a stunning success while everyday beliefs were neither better nor worse than they had been for the previous 2,000 years. The now mighty science had become worthy of philosophical attention in its own right.

So, this is where our story of key thinkers begins, roughly a century and a half ago. Laura Snyder describes the debate between Whewell and Mill, which she sees as a debate intimately linked to the social views each held. They certainly quarrelled over the nature of theories, evidence and so on. But both were also social reformers, Mill especially so. There was some interplay between their accounts of science and their views on social life. Whewell, for instance, thought some aspects of science could be known a priori. Mill took this to be based on intuition, which he thought to be nothing more that a collection of prejudices reinforcing the social status quo. His radical empiricism would serve as an antidote, both socially and scientifically.

Realism, as I mentioned above, is the commonsense view that science typically gives us a true (or approximately true) description of reality, both observable reality and the hidden entities and processes that explain the things we observe. A number of philosophers and scientists have felt compelled to reject realism and adopt some other view. Three of the most important conventionalists are Henri Poincaré, Pierre Duhem and Hans Reichenbach. They claimed that some important features of our theories should not be understood as true descriptions but rather as conventional choices. Conventionalism is a close cousin of instrumentalism; in the case of Duhem, they amount to the same thing. Torsten Wilholt explains how such a view arose and why it was and still is often maintained. Duhem was concerned with theories in general, claiming that the only thing they need do is 'save the phenomena'; that is, they should imply true observations. Like other instrumentalists, he held that the truth of theoretical assertions is irrelevant. We choose among rival theories that are equally good at saving the phenomena on the basis of convention, not truth. The choice of convention would often be determined by practical concerns, such as ease of use. Poincaré and Reichenbach largely focused on geometry, arguing that we have to make conventional choices, for instance, about what physical things might represent straight lines, and only then can we determine whether we live in a Euclidean or a non-Euclidean world. Much of conventionalism rests on what is now known as the 'underdetermination of theory by data'. The general claim is that no matter how many observations have been made, they will never be enough to pick out a unique theory; there will always be indefinitely many rival theories that are compatible with all the observations that have or ever will be made. It is a theme that comes up over and over again in the philosophy of science.

The Vienna Circle has been the inspiration or the bête noire of all subsequent philosophy of science. So many stupendously important people were involved, making Friedrich Stadler's task of describing it quite challenging. Moritz Schlick, Hans Hahn, Otto Neurath, Rudolf Carnap, Phillip Frank and many others are deeply responsible for shaping the outlook on science we have today. These are the logical positivists. Their heroes were Mach, Einstein, Russell and, for some, Wittgenstein. They added the new logic developed by Russell and others to empiricism, hence the 'logical' in logical positivism.

Most were trained in the sciences and, of course, were deeply interested in how the sciences worked. Many were also deeply political, a fact that is perhaps unsurprising given the times they lived in. All were staunch anti-Nazis; some were in the liberal centre, many were socialists or at least social democrats and others were Marxists. In some contemporary discussions the logical positivists are portrayed as right wing, which is either an expression of appalling ignorance or an outright lie.

As Stadler shows, they were also not monolithic in their views about science. Carnap and Neurath, who were friends and left wing allies, differed considerably on some aspects of the epistemology of science. And in turn, they differed from Schlick, who politically was a liberal social democrat. Schlick was murdered by a deranged student in 1936. The others left Austria (some philosophical allies, such as Reichenbach and Hempel left Germany) with the rise of the Nazis, many coming to the USA.

The influence of these émigrés was enormous, evolving into what is often called logical empiricism. Martin Curd describes the work of one of the most important of them, Carl Hempel. Along with Carnap, Reichenbach, and others, Hempel developed what came to be known as the 'received view'. This included accounts of explanation, confirmation and the structure of theories. It tended to be anti-realist about theoretical entities (e.g., 'electron' is a useful concept for organizing experience but should not be taken seriously as a real thing). It embraced a sharp fact-value distinction: science discovers the facts and society decides how to use them. And it thought there could be no such thing as a method of having good ideas, that is, there is no logic of discovery. All we can do is test conjectures, which could come from anywhere. The logical empiricist outlook dominated Anglo-American philosophy of science from the end of World War II well into the 1970s. There are still strong influences today.

Throughout the long reign of logical positivism and logical empiricism there was a strong minority view. Karl Popper, a fellow Viennese, opposed many of the Vienna Circle's main doctrines but, like them, fled Austria with the rise of the Nazis. Popper today is very well known for two things. He has been widely embraced in European political circles for his anti-Marxist philosophy. In science circles he is best known for his doctrine of falsifiability. The idea is simple, but quite profound if put into practice. It is not possible, says Popper, to confirm a theory, but it is possible to refute it. A single counter-example will do. Consequently, the practice of science should be that of making conjectures and then trying to overturn them. Any theory that does not stick its neck out, that does not make testable predictions that could turn out false, is not genuine science. Steve Fuller describes Popper's work and makes the case for his enduring importance.

Among working scientists Popper may be the best-known philosopher, but Thomas Kuhn is surely a very close rival. Kuhn, as Martin Carrier makes clear, is famous for 'paradigms'. There are many meanings of the word *paradigm*, but one is central: a specific example of a solved problem. This example is copied in using a theory in new applications; it shapes all sorts of things, including how we see the world (as a duck or as a rabbit); it implicitly provides methods of evaluation and even the very meaning of words which, Kuhn famously claimed, are incommensurable from one theory to another. Kuhn's influence has not only been enormous in the philosophy of science, it has been significant throughout all academic fields and even in general culture, where the term *paradigm* is now a commonplace.

Kuhn was perhaps the biggest influence on a movement known as HPS (history and philosophy of science), which took history and philosophy to interact with each other in a very deep way. Among the leading practitioners were Paul Feyerabend and Imre Lakatos. Though Kuhn, Feyerabend, and Lakatos drew very different philosophical morals from the history, they all would have concurred with Lakatos's famous dictum: 'History of science without philosophy of science is blind; philosophy of science without history of science is empty.'

We often think of science and its philosophy as products of the West. To a large extent this is true, but it fails to do justice to the actual situation. The fact is that huge contributions were made by Indian, Chinese, Arabic and other non-Western societies. It takes only a brief glance at Joseph Needham's work on the history of Chinese science to be disabused of the notion that science is uniquely the product of the West. (A parallel idea that democracy is a Western invention is similarly silly.) Arun Bala describes the work of Ghazali, the eleventh-century Arabic philosopher whose work on epistemology did so much for Arabic science. He was also very important as a mediator between the Greeks and the European medieval philosophers and scientists. I realize, of course, that a chapter like this is a mere token. But it may usefully remind readers that there is a much larger world out there with very many different but rich traditions.

Martin Kusch introduces us to the cluster of ideas variously known as social constructivism and the sociology of scientific knowledge. David

Bloor, Harry Collins and Bruno Latour are usually lumped together in this camp, even though they hold very different views. What unites them is the belief that a great deal of scientific activity is driven by social factors. Constructivists are often at loggerheads with mainstream philosophers of science, since the latter typically hold that evidence is the driving force, not social, political, or other non-evidential considerations. Often the debates can be quite heated, with philosophers calling constructivists empty-headed relativists, and constructivists calling mainstream philosophers naive propagandists for science. Kusch provides a very clear outline of three of the most important constructivists, allowing us to see some of the finer points of conflict.

Stathis Psillos takes on the topic of realism, truly one of the eternal topics in the philosophy of science. As I mentioned earlier, anti-realist sentiment has a long history. It was certainly a typical part of logical empiricism, the received view, around mid-twentieth century. Hilary Putnam and others argued rather successfully against this outlook and in favour of scientific realism. In reaction, Bas van Fraassen made the case for his so-called constructive empiricism, a view that distinguished between *belief* and *acceptance*. All we should do, said van Fraassen, is accept a theory, not believe it. Acceptance means only that we take it to be empirically adequate, that is, its empirical claims are true but we remain agnostic about its theoretical claims.

The realism debate explained by Psillos was a debate about theories: should we think of them as being true? A major shift was introduced in the 1980s with the rather sudden growth of interest in experiments. Nancy Cartwright, Ian Hacking, Peter Galison, Alan Franklin and several others stressed the importance of experimentation. As Hacking once put it, 'Experiments have a life of their own'. This, as William Seager makes clear, led to a new range of considerations and a new type of argument for a new type of realism, sometimes called 'entity realism'. Instead of worrying whether the theory of electrons is true, one might have direct evidence for the electrons themselves, based on the ability to manipulate them. We can have confidence in such entities, even if we do not have any confidence, say, in quantum electrodynamics, the theory that describes them. The rise of interest in experiment and the subsequent decline in theory as the principle object of philosophical interest has been one of the chief developments of recent philosophy of science.

Feminism has been one of the most important developments in philosophy of science in recent times. Needless to say, it has been and still is much resisted, but its influence has been internalized to a great extent. It is now widely recognized that the background, especially the gender, of the theorizer can be highly important to the kinds of theories produced. One of the consequences is that older notions of a value-free science have been tossed out the window. Now all sorts of factors are acknowledged as playing a role in science, though strong differences remain about the nature of this role and about what objectivity is in this new setting. Janet Kourany describes the work of two of the most important and influential feminist philosophers of science, Sandra Harding and Helen Longino. Their impact on our understanding of science, while controversial, has been and will continue to be highly significant.

A volume such as this is a compromise. Many of the key thinkers discussed are justly famous for work on several different topics, but each is singled out for only one. And the accounts of their work are not as detailed as they might be. Fuller accounts were sacrificed to make room for background and context. And, inevitably, several great philosophers have been passed over completely. I will come back to this point again in the Afterword. I hope and trust that readers will understand the inevitable space constraints and that they will enjoy what is on offer and learn from it. And now, on to the key thinkers in the philosophy of science.

CHAPTER 1

EXPERIENCE AND NECESSITY: THE MILL-WHEWELL DEBATE

Laura J. Snyder

But a still greater cause of satisfaction to me from receiving your note is that it gives me the opportunity on which, without impertinent intrusion, I may express to you how strongly I have felt drawn to you by what I have heard of your sentiments respecting the American struggle, now drawing to a close, between liberty and slavery, and between legal government and a rebellion without justification or excuse. No question of our time has been such a touchstone of men – has so tested their sterling qualities of mind and heart, as this one: and I shall all my life feel united by a sort of special tie with those, whether personally known to me or not, who have been faithful when so many were faithless.[1]

On his 71st birthday, the last he would live to see, William Whewell received his only letter from John Stuart Mill. The two men had been engaged for decades in a debate conducted in the pages of their respective books, essays, and reviews – a debate so vitriolic that Mill refused even to meet his antagonist, although Mill's closest friend, James Garth Marshall, had become Whewell's brother-in-law. Mill was finally moved to correspond with Whewell by learning that Whewell, then Master of Trinity College, Cambridge, had barred the *Times* from the Master's Lodge because of its support of the pro-slavery Southern states in the Civil War raging in America. Mill was delighted to hear that his opponent agreed with his conviction that the anti-slavery forces in the North should be supported by the British public.

It is fitting, and not coincidental, that Mill would be motivated to write to Whewell after finding out that Whewell shared his position on the most pressing moral and political issue of the age. Today, the controversy between these two exemplars of Victorian polymathy is known by philosophers primarily for its concern with issues in the philosophy of science. Indeed, it could plausibly be contended that the Whewell-Mill debate on scientific method is one of the first robust confrontations between antagonists on the topics of induction and confirmation. However, as I have argued at length in *Reforming Philosophy: A Victorian Debate on Science and Society*, the debate between Whewell and Mill crucially concerned not only science but also moral philosophy, economics, and political thought – and, moreover, their controversy on scientific reasoning cannot fully be understood without situating this controversy within a broader disagreement on these other issues. Thus examining the Mill-Whewell debate has value not only for the history of philosophy of science, but also for the study of the ways in which scientific discussions are frequently enmeshed with political and social values, whether or not that connection is explicitly stated or recognized by its participants.

The debaters[2]

William Whewell (1794–1866) was the eldest son of a master carpenter in Lancashire, in the north of England Although it was expected that he would eventually take over his father's business, his prodigious intellectual talents were discovered by the headmaster of the local grammar school, who convinced his father to allow Whewell to attend the school. Whewell later won a scholarship to attend Cambridge. He 'went up' to the University in 1812, and, professionally speaking, never left. Over the course of his long career Whewell was a lecturer in mathematics, a professor of mineralogy and a professor of moral philosophy. In 1841 he was appointed master of Trinity College, a position he held until his death twenty-five years later. Whewell engaged in his own scientific research, winning a gold medal from the Royal Society for organizing and conducting an international research project on the tides. He invented the word *scientist*, as well as numerous other new terms, including *ion*, *cathode* and *electrode* in electricity research and *uniformitarianism, catastrophism, Eocene* and *Miocene* in geology. His greatest contributions to science, however, may have been made by his writings on the history and philosophy of science and his role as teacher, mentor and friend to a generation of men who made the most important scientific discoveries of the day, including Michael Faraday, Charles Darwin, James Clerk Maxwell, James David Forbes, John Herschel and others. In addition to his works on mathematics and natural science, Whewell wrote about economics, theology, morality, architecture and international law, translated Plato and German literature and composed poetry.

John Stuart Mill (1806–1873) was the son of the philosopher James Mill and a protégé of his father's friend Jeremy Bentham. His early history as a child prodigy is well known: Greek at three, Latin at eight, etc. This was followed by his equally famous nervous breakdown as a young man, triggered in part just because of the strain of his father's and Bentham's expectations for him, and in part by his realization that he could not fully accept their utilitarian view of moral and political philosophy. In particular, he felt that both his father and Bentham ignored the importance of educating people both intellectually and morally, in order to create cultivated beings worthy of a democratic political system. Before the political system could be reformed, Mill came to believe, there had to be a 'complete renovation of the human mind' (or, as he put it with less reticence in a draft of his *Autobiography,* the 'uncultivated herd').[3] Although Mill is one of the canonical authors studied at university today, he never attended any school nor held any university post. Instead, he worked at the East India Office with his father, taking over his father's position when James Mill died. In later years he served a term in Parliament. Mill's works spanned logic, moral philosophy, economics, politics and the classics.

The debate: reforming philosophy

Both Mill and Whewell believed that in order to bring about needed reforms in the social and political realm, it was necessary to revamp philosophy. In his *Autobiography*, Mill noted that 'from the winter of 1821, when I first read Bentham . . . I had what might truly be called an object in life; to be a reformer of the world.'[4] Later, in one of his essays on

Tocqueville, Mill explained that 'economic and social changes, though among the greatest, are not the only forces which shape the course of our species; ideas are not always the mere signs and effects of social circumstances, they are themselves a power in history.'[5] And, more simply, in a letter to a friend, he wrote: 'There never was a time when ideas went for more in human affairs than they do now.'[6]

Whewell similarly felt that reforming philosophy was necessary for changing the world. In one letter, Whewell told a close friend, 'We are no longer young men. . . . I have yet to make out my case by reforming the philosophy of the age, which I am going to set about in reality. I dare say you laugh at my conceit, but you and I are friends too old and intimate I hope for me to mind that.'[7] Later, to another friend, he insisted, 'I believe we want such [new philosophical] systems more than anything else, because at the root of all improved national life must be a steady conviction of the reason, and the reason cannot acquiesce in what is not coherent, that is, systematic.'[8]

Thus, both Mill and Whewell agreed that society needed reforming and that a new political and social system could be created by refashioning philosophical systems. But disagreements over what the reformed society should look like, and how philosophy ought to be renovated to bring about this new social order, led to a long and very fruitful debate between them covering science, morality, politics and economics. It might seem surprising to find science on this list. Mill and Whewell both believed that a new philosophy of science could be employed for the purpose of transforming society. Mill was particularly explicit in noting that his philosophy of science – and, especially, his epistemology – was designed for this purpose. Mill stated this quite clearly in his *Autobiography*:

> The notion that truths external to the mind may be known by intuition or consciousness, independently of observations and experience, is, I am persuaded, in these times, the great intellectual support of false doctrines and bad institutions. . . . There never was such an instrument devised for consecrating all deep seated prejudices. And the chief strength of this false philosophy in morals, politics and religion, lies in the appeal which it is accustomed to make to the evidence of mathematics and of the cognate branches of physical science. To expel it from these, is to drive it from its stronghold. . . . In attempting to clear up the real nature of the evidence of mathematical and physical truths, the 'System of Logic' met the intuition philosophers on grounds on which they had previously been deemed unassailable.[9]

Mill designed his epistemology and philosophy of science expressly for the purpose of serving his political ends. In particular, he wanted to purge philosophy of intuitionism, which he considered the greatest threat to the types of reforms he wanted to see instituted. He believed that intuitionism supported the status quo; it allowed people to argue that what they deeply believed to be true must be true, that systems which had always existed must always exist. Thus, in order to understand his position and his debate with Whewell, we must recognize this motivation. Mill's intentions were clear in his own time; Leslie Stephen reported, somewhat surprisingly perhaps to twenty-first-century philosophers of science, that the *System of Logic* became 'a kind of sacred book for students who claimed to be genuine liberals' and was considered 'the most important manifesto of Utilitarian philosophy'.[10] In a letter to Theodor Gomperz, Mill wrote, '[You] have rightly judged that, to give the cultivators of physical science the theory of their own operations, was but a small part of the object of the book [*System of Logic*].'[11]

Whewell, too, believed that a proper epistemology and scientific method could improve society. As a student at Cambridge, he joined his close friends John Herschel, Richard Jones and Charles Babbage in a Philosophical Breakfast Club, which met on Sunday mornings after the compulsory chapel service to discuss the need for a new revolution in science.[12] This revolution was to be inspired mainly by the writings of Francis Bacon, the seventeenth-century philosopher and Lord Chancellor.

One of the main precepts of Bacon that these four friends desired to promote in their own work was his claim that science should be aimed toward the 'relief of man's estate', or the good of the common person. This aim was so important to Bacon that he told his uncle he preferred the title of 'philanthropist' to that of 'philosopher'.[13] The members of the Philosophical Breakfast Club felt that science had moved away from Bacon's view that helping to bring about the common good was the ultimate goal of inquiry, and they pledged themselves to putting science back on the right track.

Like Bacon, Whewell and his friends sought a 'logic that embraces everything', that is, an epistemology that could be applied to all areas of thought. Whewell wrote admiringly of Bacon that 'it cannot be denied that the commanding position which Bacon occupies in men's estimation arises from his proclaiming a reform in philosophy of so comprehensive a nature; a reform, which was to infuse a new spirit into every part of knowledge.'[14] This was also Whewell's project. Whewell believed that his 'antithetical epistemology', which combined empirical and idealist elements, could be useful in the battle against the dominant utilitarian moral philosophy of the Mills and Jeremy Bentham – which he felt was actually immoral and would lead to devastating social consequences if instituted widely. His philosophy of science would also be aimed against the dominant system of economics – the deductive political economy of David Ricardo and his followers – which had already led to social conclusions Whewell found repugnant, such as the passage of the New Poor Law, which forced poor labourers into workhouses.[15]

Whewell's philosophy of science

Of all the members of the Philosophical Breakfast Club, Whewell was the most astute reader of Bacon's epistemology. Like Bacon's, Whewell's epistemology was empirical, based on slow, careful generalizations from observations. Yet unlike some readers of Bacon (such as his friend Herschel),[16] Whewell recognized that Bacon also admitted that there was a rational, non-empirical ingredient to knowledge of nature; Bacon's ideal epistemologist was neither the ant, who merely gathers facts, nor the spider, who spins theoretical webs out of his own substance, but rather the bee, who takes the pollen from the flowers but then digests it and creates something different, and more palatable.[17] As Bacon himself noted, he had attempted to bring about a 'marriage' between the rational and empirical faculties.[18] The idea of consummating this marriage appealed to Whewell, who was not only a skilled experimentalist (in crystallography, mineralogy and the study of the tides) but also a talented student of Newtonian mathematical physics, with its rationalist elements of necessity and universality. Eventually, Whewell developed an 'antithetical' epistemology that combined empirical and rational parts. Every act of knowledge, according to Whewell, involves both sensations and ideas, an external and an internal element. His epistemology was, and was meant to be, a 'middle way' – like Bacon's – between pure empiricism and stark rationalism.

This epistemology is based upon the premise that our minds contain the 'germs' of certain fundamental ideas and their conceptions and that these are used in gaining knowledge of nature. Contrary to what many scholars have written, Whewell was not a Kantian epistemologist, though he did read Kant and appreciated certain elements of Kant's philosophy. One important difference is that Whewell was a realist about scientific knowledge: he believed that with the ideas and conceptions found in our minds we can have knowledge not only of our conceptually mediated experience but of the really existing physical world. Indeed, Whewell criticized Kant for viewing reality itself as a 'dim and unknown region'.[19]

Whewell was able to explain how ideas in our minds could successfully be employed to gain realist knowledge of nature by asserting a theological foundation to his epistemology: the germs of our ideas have a divine origin, he argued. Because our ideas are given to us by the Creator of the universe, who used those same ideas in His creation, we can use those ideas to gain real knowledge of the natural world; that is the very reason why we were created with the germs of those ideas. As Whewell argued with his friend, Hugh James Rose, it seemed obvious to him that there was nothing in the pursuit of scientific knowledge that could be considered damaging to a religious outlook. 'If I were not so persuaded,' Whewell admitted, 'I should be much puzzled to account for our being invested, as we so amply are, with the faculties that lead us to the discovery of scientific truth. It would be strange if our Creator should be found urging us on in a career which tended to a forgetfulness of him.'[20]

Yet Whewell claimed that before we could use these ideas to have knowledge of that world, the ideas and their conceptions must be 'explicated', or clarified and made explicit. This requires experience and experimentation as well as discussion among scientists; thus, there is an inevitably social element to knowledge. An explicated conception can be 'superinduced' onto a set of facts, combining them into an intelligible and ordered whole; Whewell calls this process 'colligation'. Whewell compares this to the act of stringing pearls onto a thread: the pearls are the facts, the thread is the conception and the pearl necklace is the law. Every empirical law thus has a rational element.

Whewell perceived that colligation is often the most difficult part of the discovery process. To find the correct conception to superinduce upon the facts is not a matter of mere guesswork. Rather, it involves different forms of inference. For example, Johannes Kepler inferred that the conception of an ellipse was the proper one to use in colligating the observed positions of the orbit of Mars; he thus discovered the law that 'the orbit of Mars is elliptical' and generalized it to 'all orbital paths are elliptical' (known now as Kepler's first law). Kepler used the observations that had been made earlier by Tycho Brahe, yet neither Tycho nor anyone else had realized that the observations could be brought together using the conception of an ellipse. It was not an accident that Kepler was able to see this: his knowledge of mathematics, and especially geometry, was superb. It was not a lucky guess. Nor was it a merely trivial process of curve fitting (as Mill later suggested). Instead, as Whewell put it, 'we have seen how long and how hard Kepler laboured, before he converted the formula for the planetary motion from an epicyclical combination, to a simple ellipse.'[21]

Once we have an empirical law, Whewell claimed, we can use three confirmation criteria to test the law: prediction, consilience, and coherence.[22] Predictive success occurs when a consequence of the law turns out to be true – especially when the consequence is a novel one, which had not been known to be true before. Consilience occurs when different inductions 'leap to the same point' or law, such as when Newton's examination of the earth's moon, the moons of Jupiter, the planets' orbits around the sun, and falling bodies all led to an inverse-square law of attraction between the involved bodies. And coherence occurs when a theory is able to take account of new phenomenal laws, or types of evidence that had not been used in constructing the original theory, as Newton's theory of universal gravitation was able to do for the laws of the tides.

If the law passes these tests, we can be extremely confident that we have a true law; at times Whewell even goes too far and suggests that we can be absolutely certain that the law is true. If it fails the tests, then either we are wrong about some of the facts, the conception is the wrong one to apply to the facts, the conception is not yet adequately explicated, or it has not been correctly applied to the facts. We must use the tests to help us refine the conception, the facts and the colligation. Discovery and confirmation are thus 'bootstrapping processes', in the sense that we go back and forth between them, using each step in the process to help modify the next.

Whewell stressed that in discovering laws by this method – which he clearly and correctly saw as an inductive method – all types of inference can be used; like Bacon, he believed that it was not enough to use only enumerative induction (of the type 'this crow is black, and this one, and this one; therefore probably all crows are black'), but also other forms of reasoning. In a series of book reviews and longer works in the 1830s, and then in his book on the 'plurality of worlds' issue in 1853, Whewell stressed the importance of analogical inference; that is, arguing from analogy. Whewell believed that science can – and indeed must – seek unobservable causes for observed regularities and recognized that analogical inference is quite useful in getting to unobservable causes, as Bacon had earlier seen.[23]

Interestingly, Whewell insisted as well that empirical science could reach necessary truths. Necessary truths are a priori truths, laws that 'can be seen to be true by a pure act of thought',[24] while empirical truths require experience to be known.[25] Moreover, necessary truths are laws 'of which we cannot distinctly conceive the contrary', while empirical or experiential truths are such that 'for anything which we might see, might have been otherwise'.[26] These are epistemic differences, in relating to how we come to know the laws and how we view them. But Whewell also defined necessary truths as those which 'must be true' and are not merely contingently true, whether we recognize that they are necessary or not.[27]

However, just as, for Whewell, the distinction between fact and theory is only a relative one – such that facts are joined together by the use of an idea or conception to form a theory, but a true theory is itself a fact and can be used to form theories of even greater generality – so too is the distinction between an empirical and necessary truth relative, and shifts over time.[28] This is what Whewell termed the 'Fundamental Antithesis' of knowledge. Like facts and theories, Whewell claimed, experiential and necessary truths 'are not marked by separate and prominent features of difference, but only by their present opposition, which is only a transient relation'.[29] Science consists in what Whewell called the 'Idealization of Facts', whereby experiential truths are 'transferred to the side' of necessary truths.[30] By this process, 'a posteriori truths become a priori truths'.[31] Interestingly, then, truths which are first knowable only empirically become knowable a priori; that is, self-evident truths *become* self-evident.

What could Whewell have possibly meant by this? Whewell claimed that necessary truths – the 'axioms' of science – can be known a priori from the Fundamental Ideas of science, because the axioms are 'necessary consequences' of these Ideas.[32] Elsewhere, he put it slightly differently, noting that the axioms express the 'meaning' of the Idea. The axioms, Whewell noted, 'in expressing the primary developments of a Fundamental Idea, do in fact express the Idea, so far as its expression in words forms part of our science.'[33] So, for instance, one of the axioms, or necessary truths, is that 'every event must have a cause'. This follows from the Idea of Cause, and indeed the axiom 'expresses, to a certain extent [because it is only one of the axioms of Cause] our Idea of Cause.'[34]

This sheds light on the progressive nature of necessary truth in science. Whewell believed that a crucial part of science is the 'explication' of Ideas and their conceptions. By this process scientists gain an explicit, more clearly grasped view of the meaning of an Idea – it becomes 'distinct'. Once an Idea is distinct enough that its meaning is understood, the scientist can see that the axioms are necessary consequences of the Idea, by virtue of the fact that they express part of this meaning. When the scientist has an explicated form of the Idea of Cause, he will know that it must be true that every event has a cause.

Thus, once an Idea is distinct, such that its meaning is clear to us, truths which we may, in fact, have discovered empirically are seen to follow a priori from the meaning of the Idea. The empirical truth becomes knowable a priori from the now-understood meaning of the Idea; the experiential truth as been 'idealized' into a necessary truth. The a priori intuition of necessary truths is 'progressive,' then, because our Ideas must be explicated before it is possible for us to know their axioms a priori. Whewell noted, for example, that 'though the discovery of the First Law of Motion was made, historically speaking, by means of experiment, we have now attained a point of view in which we see that it might have been certainly known to be true independently of experience.'[35] The First Law of Motion – which states that every object in a state of uniform motion tends to remain in that state of motion unless an external force is applied to it – is, on Whewell's view, a necessary truth that has undergone the process of idealization: though it was first knowable only by experience, it has become knowable a priori.

However, since necessary truths are those that 'must be true', regardless of our epistemic relation to them, an interesting consequence of Whewell's view emerges. He cannot, and does not, mean that truths shift from being contingent to being such that they 'must be true'. Rather, *all* laws of nature are necessary truths, in the non-epistemic sense, for Whewell, in that they follow necessarily from some Fundamental Idea used by God in creating the universe. All natural laws, therefore, are necessarily true. Whewell thus abrogated the distinction between natural law and mathematical truth.[36] In order truly to understand the natural law, the scientist must not only know that an empirical law *is* true but that it *must be* true; and he or she can understand this only by seeing how the law follows from an Idea in the mind of the Creator. That gaining this understanding is a crucial part of the task of the scientist is conveyed by Whewell's exhortation that 'Science *is* the Idealization of Facts'.[37]

Mill's philosophy of science

After this brief survey of Whewell's philosophy of science, it will not be surprising to learn that Mill saw Whewell as a purveyor of the intuitionist philosophy par excellence. Mill focused his wrath on two elements of Whewell's philosophy: the assertion that there are necessary truths, and the claim that certain conceptions provided by the mind are involved in our knowledge of the external world. In his review of one of Whewell's books on moral philosophy, the *Elements of Morality*, Mill explained that 'we do not say the intention, but certainly the tendency, of [Whewell's] efforts, is to shape the whole of philosophy, *physical as well as moral*, into a form adapted to serve as a support and justification to any opinions which happen to be established.'[38] He was pleased, though, that Whewell's *Philosophy of the Inductive Sciences* appeared as he was finishing up the *System of Logic*, because it gave him a natural target against which to aim his opposing philosophy.

The epistemology that Mill developed in opposition to intuitionism was an ultra-empiricist one, grounded solely in experience. Mill wanted to show that the only materials of our knowledge are particular, observed phenomena, and that science proceeds by generalizing these pieces of data into empirical laws. Any suggestion – such as

Whewell's – that the mind provides a conception to these phenomena was considered by Mill illegitimate, a form of intuitionism, leading inexorably to conservatism of thought (and practice). The same was true, on Mill's view, for the extension of knowledge beyond the realm of the observable – for example, to any unobservable entity postulated as the cause of observed phenomena, or to any claim of 'necessity.' As Mill clearly described his view, 'Sensation, and the mind's consciousness of its own acts, are not only the exclusive sources, but the sole materials of our knowledge.'[39]

In accordance with this position, Mill rejected necessity, both the necessary connections associated with causation and the necessary truths of science, mathematics and logic. Since necessary truths or necessary causal links cannot be known purely by experience, Mill viewed these as entailing some form of the hated intuitionism. Mill argued instead that propositions appearing to be necessary truths deriving from innate ideas are actually gained empirically. This was the case, Mill claimed, for the 'laws of thought' that constrain logic, such as the principles of contradiction and excluded middle.[40] He similarly argued that the notion of the uniformity of nature and the principle that for every event there exists a cause – presuppositions of all successful inductive reasoning – originate in experience.[41] We can see how far Mill was willing to go in his rejection of intuitionism by his denial of necessity in mathematics; he even attacked Locke's 'equivocation' in allowing that there might be necessary truths of mathematics.[42]

Causation was defined by Mill in such a way as to rule out any sort of necessary relation between the 'cause' and the 'effect'; causation, he argued, is merely the constant conjunction of observed phenomena, the empirical fact that a certain type of antecedent is invariably observed to be followed by a certain type of consequent. He announced this at the start of Book 3 of the *System of Logic*, which he prefaced with a quotation from Dugald Stewart's *Elements of the Philosophy of the Human Mind*:

> The highest, or rather the only proper objects of physics, is to ascertain those established conjunctions of successive events, which constitute the order of the universe; to record the phenomena which it exhibits to our observations, or which it discloses to our experiments; and to refer these phenomena to their general laws.[43]

Mill thus aligned himself with Stewart's position that causation was nothing but the constant connection between observed events. Mill admitted that most 'metaphysicians' argue that the notion of a cause implies something more, a 'mysterious and most powerful tie' between the cause and effect. Because of this they suppose the need for 'ascending higher, into the essences and inherent constitution of things, to find the true cause, the cause which is not only followed by, but actually produces, the effect.' Yet Mill denied that he was interested in this type of cause, even if it were possible to reach it. Rather, Mill explained that his scientific epistemology sought only 'physical causes', that is, the observed antecedent that is invariably followed by an observed consequent.[44] As he put it, 'when in the course of this inquiry I speak of the cause of any phenomenon, I do not mean a cause which is not itself a phenomenon.'[45] In a later work, *Auguste Comte and Positivism*, Mill even more explicitly defined the law of universal causation as stating that 'every phenomenon has a phenomenal cause'.[46]

Mill's rejection of necessary connection in causation was motivated not only by his general worries about intuitionism, but more specifically by his concern to allow a certain kind of human freedom in the moral sphere.[47] Mill himself drew this connection between the topic of causation and moral freedom in a long footnote to the discussion of the law of causation in Book 3 of the *Logic*, and developed it further in his chapter 'Of Liberty and Necessity' in Book 6. Mill argued that the law of causality applies to human actions as it does to other natural phenomena. As Mill noted, many people reject this position – called 'determinism', or 'philosophical necessity' – on the grounds that it implies we are forced to act against our will, or without control over our actions. However, Mill pointed out, this implication follows only from an incorrect concept of causation, one that holds there is some 'mysterious constraint exercised by the antecedent over the consequent'. Once the correct notion of causation is applied to the realm of moral action, there is no such problem. As Mill put it,

> Those who think that causes draw their effects after them by a mystical tie, are right in believing that the relation between volitions and their antecedents is of another nature. But they should go farther, and admit that this is also true of all other effects and their antecedents. If such a tie is considered to be involved in the word necessity, the doctrine is not true of human actions; but neither is it

then true of inanimate objects. It would be more correct to say that matter is not bound by necessity, than that the mind is so.[48]

We do not feel that our own 'freedom' would be degraded by the doctrine that 'our volitions and actions are invariable consequents of our antecedent states of mind.' Such a position does not suggest that a person's actions are compelled or necessitated by either his character or his circumstances; if he wished to resist a certain motive, he could do so – because that wish to resist would be one further antecedent state of mind. Mill is in this way reducing philosophical 'necessity' to predictability, rather than compulsion. If a person's character and 'all inducements acting upon' him at a given moment were known, his actions could be predicted reliably.

What Mill wished to accomplish by proposing this view of human causation was to demonstrate the importance of moral self-education. The person who does something wrong is not compelled by any of her desires to do so; rather, she does not control her desires properly. A moral education strengthens the desire for what is right and the aversion to what is wrong, so that we will desire and choose the right actions. Mill was here countering the claim of the social reformer Robert Owen and his followers, who were then arguing that individuals should not be punished for their wrong actions, because they cannot help how they act; their actions are necessary consequences of their characters, which are not under their control. Mill, however, argued that men and women are responsible for their actions, because they have the power (and the responsibility) to alter their characters – by providing themselves with the proper kind of moral education. 'Not only our conduct, but our character, is in part amenable to our will', Mill asserted.[49] If we do not form our own character, by choosing our own desires and impulses, we can be said to have no real character. Using a timely example, Mill derisively stated, 'One whose desires and impulses are not his own, has not character, no more than a steam-engine has a character.'[50] Indeed, Mill defined moral freedom as the ability to modify our characters as we wish.[51] On Mill's view, the very possibility of moral freedom required his notion of causation, that which rejected non-phenomenal causes.

Further, Mill rejected the appeal to any unobservable entities or properties in science, claiming that we cannot have any knowledge of these.

Although in later editions Mill toned down this comment, he succinctly expressed his reason for rejecting the hypothesis of the 'luminiferous aether' postulated by adherents of the wave theory of optics because 'it could never be brought to the test of observation, because the ether is supposed wanting in all the properties by means of which our senses take cognizance of external phenomena. It can neither be seen, heard, smelt, tasted, nor touched.'[52] Mill did allow that unobserved causes may be postulated 'hypothetically', but clearly noted that they remain hypothetical until 'brought to the test of observation'. In later editions of the *Logic*, Mill would point to Darwin's theory of evolution by natural selection as a cause postulated only hypothetically, which still awaited being successfully 'brought to the test of observation'. (Thus Mill, completely misunderstanding Darwin's *Origin of Species*, wrote that 'Mr. Darwin has never pretended that his doctrine was proved.')[53]

Indeed, in his *Examination of Sir William Hamilton's Philosophy*, Mill suggested another reason for rejecting the appeal to unobserved causes in science: he did not admit the very existence of such causes. In this work on metaphysics, Mill admitted that he was following his father and David Hartley in their associationist way of characterizing 'matter' as only the 'permanent possibilities of sensation'. Mill thus implied that he rejected the very existence of entities that could invoke no sensations. Indeed, Mill himself addressed the meaning of this phrase, writing 'matter . . . may be defined [solely as] a Permanent Possibility of Sensation. . . . If I am asked, whether I believe in matter, I ask whether the questioner accepts this definition of it. If he does, I believe in matter, and so do all Berkeleians. . . . In any other sense than this, I do not.'[54] Like Berkeley, Mill rejected the existence of anything besides sensations or ideas in individual minds. The belief in a permanent, unexperienced 'substratum' of matter underlying and causing those sensations is merely an inference – rather, a hypothesis, and one that can never be 'brought to the test of observation'.[55] However, unlike Berkeley, Mill identified matter as the 'potentiality' of sensations, not the sensations actually existing in any minds (this is what led Berkeley to what Mill called the 'weak and illogical' part of his view, the invocation of a Divine Mind, in which sensations exist even if they are not being experienced by any other mind).[56]

Thus, Mill turns out to have rejected not only the possibility of probable knowledge of unobservable causes, but also their very existence.

Mill's view of matter was that it is nothing but 'the permanent possibilities of sensation' or (and he seemed to think of these phrases as equivalent) 'power[s] of exciting sensations'. Unobservable entities or properties, which cannot possibly cause sensations, therefore have a problematic status for Mill. Entities or properties that are unobserved – or even unobservable given our present location in the universe, or the abilities of our sense-enhancing instruments – can be said to exist, as long as it is possible that they could potentially exercise the power of causing sensations at a future time. But entities or properties that are 'unobservable in principle' – such as the luminiferous ether, which was considered undetectable by any of our senses, no matter how augmented by instrumentation now or in the future – cannot be said to exist, on Mill's view.

Finally, Mill rejected the notion that knowledge requires any contribution from the mind, any conceptions that originate within us. 'Conceptions do not develop themselves from within, but are impressed from without', Mill insisted. 'The conception is not furnished *by* the mind until it has been furnished *to* the mind.'[57] He disagreed explicitly and vehemently with Whewell's characterization of Kepler's discovery, insisting that Kepler either merely described the observations of the curve of the orbit of Mars, or that he used a purely hypothetical process, 'guessing until a guess is found that tallies with the facts'.[58] In neither case, according to Mill, does Kepler deserve any credit for a truly inductive discovery.

Mill's methodology is, like his epistemology, purely empiricist. It relies on his famous 'Methods of Experimental Inquiry', the methods of difference, agreement, joint method of agreement and difference, concomitant variation, and residues. Only the method of difference can reach specific causal laws – by which, remember, Mill meant only laws of constant conjunction of observed entities (by the use of the other methods we can conclude that there is some causal relation between types of events, but not the exact nature of this connection; that is, which one is the cause and which the effect). The method of difference is characterized by Mill in the following way:

A, B, C ◊ a, b, c
B, C ◊ b, c

Therefore, (probably) A is the cause of a

In order to apply the method, it is necessary to have observed entities and instances; one must actually see the event types characterized schematically as 'A, B and C', as well as those represented by 'a, b and c', in order to use this method. Thus, the methods of experimental inquiry can be used only on types of events that are not only observable in principle but actually observed.

Most cases in sciences require a more complex methodology, what Mill calls the 'deductive method'. This is necessary when there is an 'intermixture of effects'; that is, more than one cause operating at the same time (Mill likens this to the 'composition of forces' in mechanics). This method consists in a three-step process: first, an induction, using the methods of experimental inquiry; then, a deduction of the consequences of the induction; and finally a verification of these consequences. Here, too, all the involved processes and entities must be observed, in order for the first step to be conducted. Mill's philosophy of science was, consistent with his rejection of intuitionism, purely phenomenalist, ruling out the possibility of theoretical science of any kind.

Whewell's response to Mill

It was not until six years after the publication of *System of Logic* that Whewell publicly responded (this was clearly a disappointment to Mill, who admitted that he had hoped Whewell's propensity for 'polemics' would cause him to review the work right away). In 1849, Whewell published a monograph, *Of Induction, with Especial Reference to Mr. J. Stuart Mill's 'System of Logic'*. Much of this work is taken up with criticisms of Mill's view of Kepler's discovery. For example, while agreeing that Kepler's discovery amounted to the realization that the observed points of Mars's orbit shared a 'commonality', as Mill had put it, Whewell scathingly noted that 'it appears to me a most scanty, vague and incomplete account' to suggest that that commonality – lying on an elliptical curve – was found merely by observation.[59] Whewell insisted that the fact 'that the orbit of Mars is a Fact – a true description of the path – does not make it the less a case of Induction.'[60] Indeed, all true inductions lead to facts. That some are descriptions of spatial properties and others are descriptions of causal ones seems not to make any logical difference. If we cannot directly observe the orbital path (and we cannot), then we need inference to discover what it is.

Whewell also opposed Mill's radical empiricism. Like Comte, Whewell dismissively noted, Mill 'rejected the inquiry into causes', which makes the mistake of 'secur[ing] ourselves from the poison of error by abstaining from the banquet of truth'.[61] The methods of experimental inquiry Whewell viewed as extreme oversimplifications of scientific discovery. He rejected Mill's suggestion that there can be a 'discovery machine', or an algorithm for finding new empirical laws (Mill had suggested as much by claiming that his methods were 'the only possible modes of experimental inquiry' and that they were thus analogous to the rules of the syllogism in deductive logic).[62] Finally, Whewell complained about the lack of historical justification for Mill's methodology. Mill had not shown that his methods were actually the ones used by anyone making any discoveries in past or present times. As Whewell protested, 'Who will carry these formulae through the history of the sciences, as they have really grown up; and shew us that these . . . methods have been operative in their formation?'[63] In his own aptly titled *Philosophy of the Inductive Sciences, Founded upon Their History*, Whewell had taken great pains to show that *his* view of scientific method was based on his previous study of the history of all the natural sciences. Whewell was dismayed that Mill had not taken the same pains to show that his view was supported by an inductive study of the history of science.[64] Ironically, then, Whewell's major criticism of Mill's ultra-empiricist philosophy of science was that Mill did not use the evidence of experience to justify that philosophy.

Conclusion

I have claimed that to understand fully the Mill-Whewell debate on science, it is necessary to view their dispute in the context of their broader aims of reforming society and philosophy in general. While space has not allowed a full fleshing out of this claim, I have at least shown how Mill's epistemology and scientific methodology was clearly and explicitly based on his disdain for the 'intuitionist' philosophy and shown his conception of Whewell's philosophy of science as exemplifying this view. By the end of their lives, Mill had come to see that

Whewell was not the raging political conservative that Mill had always assumed he must be, given his 'intuitionist' tendencies. Unfortunately, we will never know if this realization might have sparked Mill to rethink the political motivation for his extreme phenomenalist philosophy of science – for within a few years of his letter to Whewell, Mill too would die, leaving the Mill-Whewell debate in the hands of historians of philosophy of science.

Notes

1   Mill to Whewell, 24 May 1865, Whewell Papers, Trinity College, Cambridge (hereafter WP) Add.ms.a.209 f. 48 (1).

2   A much more detailed discussion of the Mill-Whewell debate in its entirety can be found in my *Reforming Philosophy: A Victorian Debate on Science and Society* (Chicago: University of Chicago Press, 2006).

3   For the first quote, see Mill, 'Grote's *History of Greece* (V),' in *Collected Works of John Stuart Mill* (hereafter *CW*), general editor John M. Robson, 33 vols (Toronto: University of Toronto Press; London: Routledge and Kegan Paul, 1963–91), CW 25:1162; for the second, Mill, *Early Draft of Autobiography*, CW I:112, 238.

4   Mill, *Autobiography*, CW I:137.

5   Mill, 'De Tocqueville on Democracy in America' (II), CW 18:197–8.

6   Mill to Robert Barclay Fox, 9 September 1842, CW 13:544.

7   William Whewell to Hugh James Rose, summer 1836, quoted in Janet Mary Stair Douglas, *The Life and Selections from the Correspondence of William Whewell, D.D.*, London: C. Kegan and Paul, 1882, p. 181.

8   Whewell to Julius Charles Hare, 13 March 1842, WP Add.Ms.a.215 f. 266.

9   Mill, *Autobiography*, CW 1:233–5.

10 Leslie Stephen, *The English Utilitarians*, vol. 3, *John Stuart Mill* (London: Duckworth, 1900), pp. 75–6.

11 Mill to Theodor Gomperz, 19 August 1854, CW 14:238.

12 This group and the revolution in science they succeeded in bringing about is the topic of my *The Philosophical Breakfast Club: Four Remarkable Friends Who Transformed Science and Changed the World* (New York: Broadway Books 2011).

13 For more on this aspect of Bacon's thought, see Benjamin Farrington, *Francis Bacon, Philosopher of Industrial Science* (London: Lawrence and Wishart, 1951).

14 Whewell, *On the Philosophy of Discovery, Chapters Historical and Critical* (London: John W. Parker, 1860), p. 126.

15 Unfortunately, lack of space precludes a fuller discussion of these points. For more on Whewell's opposition to utilitarianism, see my *Reforming Philosophy*, chapter 4, and on his opposition to David Ricardo's political economy, see *Reforming Philosophy*, chapter 5, and *The Philosophical Breakfast Club*, chapter 5.

16 For the different readings of Bacon by Herschel and Whewell, see *Reforming Philosophy*, pp. 80–2, and 'Bold Leaps: Guesses or Inferences? John Herschel and Analogical Reasoning in Science', forthcoming.

17 Francis Bacon, *The Works of Francis Bacon,* collected and edited by J. Spedding, R. L. Ellis, and D. D. Heath, 14 vols (London: Longman, 1857–61, new edn, 1877–89), 4:92–3. See also Paolo Rossi, 'Ants, Spiders, Epistemologists', pp. 245–60 in *Francis Bacon: Terminologia e Fortuna nel XVII Secolo*, edited by M. Fattori (Rome: Edizioni dell'Ateneo, 1984), p. 255.

18 See Francis Bacon, *Works*, 4:19.

19 Whewell, *On the Philosophy of Discovery: Chapters Historical and Critical* (London: John W. Parker, 1860), p. 312. For more on the differences between the views of Whewell and Kant, see my *Reforming Philosophy*, pp. 42–7.

20 Whewell to Rose, 19 November 1826, WP R.2.99 f. 26.

21 Whewell, *Novum Organon Renovatum* (London: John W. Parker, 1858), p. 201.

22 Lack of space prevents a full discussion of these confirmation tests. For more, see my *Reforming Philosophy*, pp. 167–202, and on consilience and its relation to realism, 'My Consilience, Confirmation, and Realism', in P. Achinstein (ed.), *Scientific Evidence: Philosophical Theories and Applications* (Baltimore: Johns Hopkins University Press, 2005), pp. 129–48.

23 Herschel also stressed the importance of analogical causes in his notion of the *vera causa*. I discuss this point in 'Bold Leaps', forthcoming.

24 Whewell, *The History of Scientific Ideas*, 2 vols (London: John W. Parker, 1858), I:60.

25 *The History of Scientific Ideas*, I:26.

26 *The History of Scientific Ideas*, I:25, and Whewell, 'On the Fundamental Antithesis of Philosophy', *Transactions of the Cambridge Philosophical Society* 8, pt II (1844): 170–81; reprinted in *On the Philosophy of Discovery* (1860) as Appendix E, pp. 462–81 (p. 463).

27 *The History of Scientific Ideas*, I:25–6.

28 See 'On the Fundamental Antithesis of Philosophy', p. 467.

29 *On the Philosophy of Discovery*, p. 305.

30 *On the Philosophy of Discovery*, p. 303.

31 *On the Philosophy of Discovery*, pp. 357–8.

32 *The History of Scientific Ideas*, I:99.

33 *The History of Scientific Ideas*, I:75, I:58, and *Novum Organon Renovatum*, p. 13.

34 *The History of Scientific Ideas*, I:185.

35 Whewell, *The Philosophy of the Inductive Sciences, Founded Upon Their History*. 2nd edn, 2 vols (London: John W. Parker, 1847), II:221.

36 The nineteenth-century logician and Kantian Henry Mansel saw this conse-
    quence of Whewell's philosophy of science and complained that the difference
    between a priori principles and empirical laws 'is not one of degree, but of kind;
    and the separation between the two classes is such that no conceivable progress
    of science can ever convert the one into the other.' H. L. Mansel, *Prolegomena
    Logica: An Inquiry into the Psychological Character of Logical Processes*, 2nd
    edn (Boston: Gould and Lincoln, 1860), p. 275.
37 Whewell, *The Philosophy of the Inductive Sciences*, I:46.
38 Mill, 'Dr. Whewell's Moral Philosophy', CW 10:168; italics added.
39 Mill, 'Coleridge', in CW 10:125.
40 Mill seemed to back down from this claim to some extent at the end of his
    life. See his 'Grote's Aristotle', CW 11:499–500, and Geoffrey Scarre, *Logic
    and Reality in the Philosophy of John Stuart Mill* (Dordrecht: Reidel, 1989),
    pp. 138–45.
41 See Scarre, *Logic and Reality in the Philosophy of John Stuart Mill*, pp. 80–7.
42 See Alan Ryan, *John Stuart Mill* (New York: Pantheon, 1970), pp. 75ff. and
    Scarre, *Logic and Reality in the Philosophy of John Stuart Mill*, p. 3. For more on
    these topics, see my *Reforming Philosophy*, pp. 106–12.
43 Mill, *System of Logic*, CW 7:282.
44 Mill recognized, of course, that mere invariability was not enough, or else night
    would be the cause of day (and vice versa). He also noted that causality requires
    'unconditionality', in the sense of not being dependent on any other anteced-
    ents. By this criterion, Mill explained, we can see that the cause of day is the
    presence of the sun over the horizon (see *System of Logic*, CW 7:335). After
    being criticized for including a criterion that could not be known solely by expe-
    rience, Mill elaborated in later editions that we *can* know this by experience; for
    example, by seeing the day come once the sun appears over the horizon every
    24 hours (*System of Logic*, CW 7:340).
45 Mill, *System of Logic*, CW 7:326–7, and see Mill, *Examination of Sir William
    Hamilton's Philosophy*, CW 9:362.
46 Mill, *Auguste Comte and Positivism*, CW 10:293.
47 For a more detailed discussion of this point, see my 'Freedom from Necessity:
    The Influence of J. S. Mill's Politics on His Concept of Causation', in P. Machamer
    and G. Wolters (eds), *Thinking About Causes: From Greek Philosophy to Modern
    Physics* (Pittsburgh: University of Pittsburgh Press, 2007), pp. 123–40.
48 Mill, *System of Logic*, CW 8:838. In Book 3, Mill wrote that 'our will causes our
    bodily actions in the same sense, and in no other, in which cold causes ice, or a
    spark causes an explosion of gunpowder' (CW 7:355).
49 Mill, *Examination of Sir William Hamilton's Philosophy*, CW 9:465–6.
50 Mill, *On Liberty*, CW 18:264.
51 Mill, *System of Logic*, CW 8:841.
52 Mill, *System of Logic*, CW 7:499. Whewell would later ridicule Mill for this com-
    ment, and in response Mill omitted this phrase from his third and later editions

of the *Logic*. But Mill continued to believe that, in the case of unobserved causes, we could not have probable knowledge of their existence, because there would be no way to rule out other conflicting hypotheses that also account for all the observed phenomena. See *Logic*, CW 7:500.

53 Mill, *System of Logic*, CW 7:498–9n. Notice that he added his discussion of Darwin to the section on 'the indispensableness of scientific hypotheses', where Mill argued that hypotheses have, basically, what we would call 'heuristic value', because they give 'inducement' to trying certain experiments rather than other ones. Of Darwin's 'hypothesis', Mill asked, 'is it not a wonderful feat of scientific knowledge and ingenuity to have rendered so bold a suggestion, which the first impulse of every one was to reject at once, admissible and discussable, even as a conjecture?' (*System of Logic*, CW 7:499n).

54 Mill, *Examination of Sir William Hamilton's Philosophy*, CW 9:187.

55 Mill explained, 'I assume only the tendency' to go beyond experience, 'but not the legitimacy of the tendency'. See *Examination of Sir William Hamilton's Philosophy*, CW 9:187.

56 See Mill, 'Berkeley's Life and Writings', CW 11:464, and *Examination of Sir William Hamilton's Philosophy*, CW 9:201.

57 Mill, *System of Logic*, CW 8:653, 655.

58 Mill, *System of Logic*, CW 7:304.

59 Whewell, *Of Induction, with Especial Reference to Mr. J. Stuart Mill's 'System of Logic'* (London: John W. Parker, 1849), pp. 41–2.

60 Whewell, *Of Induction*, p. 23.

61 Whewell, *On the Philosophy of Discovery*, p. 233; see also Whewell, 'Comte and Positivism', *Macmillan's Magazine* 13 (1866): 353–62; esp. 356.

62 Mill, *System of Logic*, CW 7:406.

63 Whewell, *Of Induction*, p. 45.

64 More on this can be found in my *Reforming Philosophy*, pp. 141–55.

CHAPTER 2

CONVENTIONALISM: POINCARÉ, DUHEM, REICHENBACH

Torsten Wilholt

Conventionalism

A recurrent theme in philosophy of science since the early twentieth century is the idea that at least some basic tenets within scientific theories ought to be understood as conventions. Various versions of this idea have come to be grouped together under the label 'conventionalism'. At least implicitly, conventionalism emphasizes the social character of the scientific enterprise. A convention, after all, is a solution to a problem of coordination between multiple actors (cf. Lewis 1969, ch. 1). Whenever achieving the best solution to a given problem depends not on every actor choosing one uniquely optimal course of action, but rather on everyone choosing the same option (or, as the case may be, corresponding ones), we say that the problem can be solved by agreeing on one approach 'by convention'. The notion that some statements of scientific theories are conventional thus goes against a (self-) understanding of science as an essentially individualistic endeavour, according to which any scientific investigation could in principle (i.e., if only it weren't too much work) be performed by any intelligent individual using the right methods and reason and experience would uniquely determine the investigation's conclusions. Deviating from this understanding, philosophers of science of the first half of the twentieth century have devised some influential arguments in support of an essential role for convention in science. The story of conventionalism has its roots in the philosophical reflection on the nature of space and the status of geometry.

Poincaré

The curious nature of geometry has taxed many philosophers. On the one hand, it is a mathematical discipline within which theorems are established by proof and without reference to experiment or observation. On the other hand, it seems to make direct claims about the physical world that we inhabit and experience. In the late eighteenth century, the philosopher Immanuel Kant offered a solution to the problem of integrating these two aspects into one coherent conception of geometry (Kant 1781/87, B37–B45). For Kant, space is a 'pure form of intuition' and belongs to the principles that the human mind provides for structuring and ordering sense experience. All our experience of external objects is structured by the spatial form of intuition, which is why our geometrical knowledge holds good for all the world of experience. However, in order to attain geometrical knowledge, we need not have any actual experience of external objects. The proofs of geometrical theorems can be constructed in pure intuition itself, thereby rendering geometrical knowledge a priori – that is, independent of sense experience. That they *have* to be thus constructed shows, according to Kant, that the truths of geometry are not just consequences of the meanings of the terms *point*, *straight line* and so forth – in Kantian terms, geometrical knowledge is not analytic, but synthetic.

Kant's conception of geometry as a body of synthetic a priori truths soon became immensely influential. But the nineteenth century saw developments within the science of geometry itself that would ultimately shake its foundations. These developments began when non-Euclidean geometries were discovered. Geometers had long puzzled over the status of Euclid's fifth postulate, the so-called parallel postulate. It is equivalent to the claim that given a straight line *l* and a point *P* not on *l*, there exists exactly one straight line through *P* that is parallel to *l*. Euclid needed the postulate to prove basic geometrical propositions, such as the theorem that the angle sum of a triangle is always equal to 180 degrees. To many geometers the postulate itself seemed less simple and self-evident than the other axioms. However, all attempts to prove it on the basis of the remaining axioms failed. In the 1820s, two mathematicians working independently – a Russian, Nikolai Lobachevsky, and a Hungarian, János Bolyai – managed to demonstrate that the parallel postulate cannot be proven from the other Euclidean axioms. In effect they showed that the other axioms are logically consistent with the assumption that more than one parallel to *l* can be drawn through *P*. Later, Bernhard Riemann showed that the other Euclidean axioms are also consistent with the assumption that the number of parallels to *l* through *P* is zero. Riemann was one of the first and most successful among those mathematicians who took up the task of exploring and systematizing the non-Euclidean geometries, in other words, the various consistent systems that can be generated by dropping the parallel postulate and replacing it with alternative assumptions.

At the same time, a second task that slowly attracted attention was that of determining the epistemological status of the various non-Euclidean geometries, especially in relation to Euclidean geometry. The Kantian position that the latter necessarily constituted the uniquely true conception of space came to be doubted. To illustrate the challenge, consider the following thought experiment, which was used in a similar fashion by the German physicist Hermann von Helmholtz in an influential paper (Helmholtz 1876). Imagine a two-dimensional world – let's call it Flatland – inhabited by two-dimensional creatures.[1] But do not imagine their universe as a flat plane, but instead as the surface of a large sphere. The creatures cannot perceive the spherical form of their world, as they are themselves two-dimensional and ought to be pictured as living within the surface rather than on it. The shortest path between two points on a spherical surface is an arc of the great circle through these two points. (A great circle is any circle on a sphere's surface that exactly cuts it in half.) So what is a straight line for the Flatlanders is, from our perspective, a great circle. In the geometry of Flatland, there is never a parallel to a given straight line *l* that goes through a point lying outside *l*, because two numerically distinct great circles on the surface of a sphere always intersect. Another curious fact about Flatland geometry is that the angles of triangles do not add up to 180 degrees. To see this, consider one of the triangular pieces of orange peel that you get by cutting an orange into eight pieces of exactly the same size by perpendicular cuts. All the cuts meet at right angles, so the angle sum of such a triangle is 270 degrees.

If you cut out smaller triangles of orange peel (always along great circles on the orange's surface), you will get angle sums of less then 270 degrees, but always more than 180 degrees. The geometry of Flatland is thus a close illustration of the two-dimensional case of that variety of non-Euclidean geometry discovered by Riemann which is now called elliptical geometry.[2] Since the Flatlanders have never known anything other than Flatland, the geometry of their school-books may very well be of the elliptical variety. To them, a geometry that claims an identical angle sum for all triangles independent of their size might seem as absurd and immediately refuted by geometrical intuition as the idea of a straight line that has no parallels seems to many humans. On the other hand, a perhaps even more thought-provoking idea is the hypothesis that the Flatlanders might in fact use Euclidean geometry and get along fine with it – provided that they inhabit (and confine all their measurements to) only a very small region of the Flatland universe. A small enough region of a spherical surface could approximate a Euclidean plane so closely that the deviations would be too small to be measured. This reveals that the situation of the Flatlanders could be our own. While it is impossible for us to visualize how our three-dimensional space could be 'curved' (just as the two-dimensional Flatlanders would perhaps have no way of pic-turing *their* universe as a curved surface), the mathematical possibil-ity that a three-dimensional space can have such structure is exactly what is proven by the existence of three-dimensional non-Euclidean geometries. Helmholtz (and others) concluded that the geometry of physical space is an empirical matter, to be decided by measurement. (Helmholtz also thought that astronomical measurements gave some evidence in favour of a Euclidean geometry of physical space.)

It was in this general context that conventionalism was first pro-posed as a philosophical viewpoint on the status of geometry, as an alternative to both Kantian apriorism and Helmholtzian empiricism. The man who made this proposal was none other than Henri Poincaré (1854–1912), a celebrated French mathematician with a staggering list of achievements in a wide range of mathematical specialties, including complex analysis, algebra, number theory and topology. In addition, he made important contributions to mathematical physics and to celes-tial mechanics.[3] Starting with a series of papers in the 1890s (many of which were later published in Poincaré 1902), he also addressed philosophical questions about mathematical and scientific knowledge, and in particular about geometry.

With Helmholtz (and also alluding to the Flatland thought experiment), Poincaré rejects the Kantian claim that our choice of geometry is a priori restricted (Poincaré 1902, 49). But he also denies that experience decides the matter. Selecting a geometry involves a real choice, where experience merely plays the role of giving 'the indications following which the choice is made' (Poincaré 1905, 72). Another thought experiment, presented by Poincaré himself in support of this conclusion (1902, 65–8), will help us to explain his reasoning.

This time, we are to imagine a three-dimensional world filling up the inside of a sphere – let's call it Sphereland. This world has an absolutely stable temperature distribution, where temperature varies only with distance r from the centre of the sphere. The law according to which it varies is $R^2 - r^2$ (where R is the radius of the sphere itself), so that temperature is highest in the centre and approaches zero toward the edge of the world. All bodies expand and contract thermally, such that their linear dimensions vary proportionally with temperature. In addition, the refractive index of the optical medium that fills this world also varies with $R^2 - r^2$. As a result, light travels not on straight lines, but on circular ones (that cut the sphere's edge at a right angle). So far, we have described Sphereland using Euclidean geometry. But how are the Spherelanders themselves likely to describe it? To begin with, they will not regard their world as confined within a sphere, but as infinite, because as they approach the sphere's surface, they (and their steps) get smaller and smaller, rendering them unable ever to reach an end of their world. They obtain the same result if they take a solid body, use it as a ruler and try to measure the diameter of the universe. (Figure 2.1 shows how the ruler contracts – from our point of view. The ruler is bent because of the stronger contraction effects on those parts of its material that are further away from the centre of the world.)

If the Spherelanders take light to travel along straight lines, they will find that given a line l and a point P not on it, more than one line exists in the same plane that passes through P and vanishes into infinity without ever intersecting l (as illustrated by lines m and n in Figure 2.1). And if they measure the angles of a large triangle with an optical surveying instrument, they will find the angle sum to be less than 180 degrees (compare the triangle PQS in Figure 2.1). As a result, the geometry that

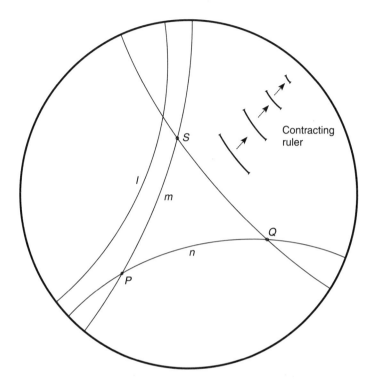

Figure 2.1 Sphereland (a planar cut through its centre).

the Spherelanders must adopt if they regard light rays as straight lines and solid bodies as reliable rulers is of a non-Euclidean kind – to be more precise, it is the 'hyperbolic' type of non-Euclidean geometry that was discovered by Lobachevsky and Bolyai. Of course, they will then *not* agree that bodies contract as they are transported away from the centre of the universe (measured with their rulers, things retain their size) nor that light rays in Sphereland are bent by a medium of varying optical density (light rays follow what Spherelanders take to be straight lines).

But what if a revolutionary Sphereland physicist were to challenge this hyperbolic world view? He might instead propose the Euclidean description that we used at the outset to introduce Sphereland – bodies contract, therefore all measurements made by the Spherelanders with rulers need to be corrected by a correction factor; likewise for angular measurements with optical instruments, because light rays are bent, and so forth. This Euclidean view of Sphereland posits both a different geometry and different physical laws about the behaviour of bodies and of light, but the overall description of Sphereland that it gives is just as accurate as the hyperbolic one. No measurement or observation could decide between the two.

Poincaré concludes that this is generally the case. If, for example, we ourselves were to measure the angles of a large triangle in the actual universe with an optical instrument and come up with a result other than 180 degrees, we would always have a choice; we could either relinquish the assumption that space is Euclidean, or we could give up the assumption that light travels along straight lines. There is nothing in the science of geometry that dictates which objects in the physical world are to be identified with straight lines and rigid bodies. Geometry in and of itself can therefore never be tested by physical measurement; it is only the conjunction of geometry with a set of assumptions about the behaviour of certain physical things (solid bodies, rays of light) – in short, the conjunction of geometry and physics – that can conform or fail to conform to empirical evidence. By changing the physics, we can uphold whichever geometry we want: '[N]o experiment will ever be in contradiction with Euclid's postulate; but, on the other hand, no experiment will ever be in contradiction with Lobatschewsky's postulate' (Poincaré 1902, 75).

Later philosophers of science would use the word 'underdetermination' to describe situations like the one analyzed by Poincaré: The choice of geometry is underdetermined by experience, because two descriptions of the world can be given that differ with regard to geometry but have exactly the same observable consequences. However, while Poincaré insisted on a real choice in connection with geometry, he did not think that the choice was an arbitrary one. Geometries differ in convenience. Which one is most convenient depends on experience, which therefore 'guides us' in our choice of geometry (Poincaré 1902, 70–1).[4] Given our actual experience with the properties of solid bodies and given the superior simplicity of Euclidean geometry, Poincaré is convinced that 'Euclidean geometry is, and will remain, the most convenient' (ibid., 50). Nonetheless, its axioms are 'neither synthetic a priori intuitions nor experimental facts. They are conventions' (ibid.).

Poincaré also took some careful and very restricted steps to extend his conventionalist thesis beyond geometry. Certain laws in the experimental sciences (in particular, in mechanics) take on such a central and entrenched place within the system of thought that they come to be regarded as principles or definitions. One of his examples is Newton's law of acceleration, $F = ma$. In the established systematic treatment of classical mechanics it is best understood as a definition of force, says Poincaré. 'It is *by definition* that force is equal to the product of the mass and the acceleration; this is a principle which is henceforth beyond the reach of any future experiment' (ibid., 104).

With regard to Newton's law of gravitation, Poincaré gives the following explanation of how a law, once elevated 'into a principle by adopting conventions' (1905, 124), can be excluded from experimental testing:

> Suppose the astronomers discover that the stars do not exactly obey Newton's law. [. . .] We can break up this proposition: (1) The stars obey Newton's law, into two others; (2) gravitation obeys Newton's law; (3) gravitation is the only force acting on the stars. In this case proposition (2) is no longer anything but a definition and is beyond the test of experiment; but then it will be on proposition (3) that this check can be exercised. (ibid.)

Even in the face of deviating observations, the law of gravitation can thus be saved from experimental disconfirmation by dropping the assumption that the behaviour of heavenly bodies is governed by gravitation alone.

The thought that some elements of scientific theory become so well established that they rise to the rank of conventions was further generalized by later philosophers of science, including Rudolf Carnap (see Chapter 3). The idea was later understood in such a way that the conventional elements of scientific theory, rather than expressing facts, span the conceptual framework within which facts can be expressed. For Poincaré, the border between conventional and empirical elements in science was not entirely strict. He explicitly emphasizes that even conventional principles can be destabilized by recalcitrant empirical evidence. In the face of continuing problems to square a principle with experience, it becomes empty and unfruitful and is thereby undermined (Poincaré 1905, 109–10).

Poincaré also emphasizes that conventionalism does not mean that scientists can devise laws as they please. 'No, scientific laws are not artificial creations' (1905, 14) – even the conventional principles were first discovered as empirical regularities. What is conventional is merely the choice of which to elevate to the rank of principles. 'Conventions, yes; arbitrary, no – they would be so if we lost sight of the experiments which led the founders of the science to adopt them, and which, imperfect as they were, were sufficient to justify their adoption' (Poincaré 1902, 110).

Poincaré's significance for the philosophy of science is not limited to the origins of the conventionalist tradition. He also made important contributions to the philosophy of arithmetic (see Folina 1992). Interestingly, there he defended a decidedly pro-Kantian view, arguing that a core principle of arithmetical proofs, the principle of complete induction, was grounded in intuition. He therefore thought that arithmetical truths were synthetic rather than analytic, as Bertrand Russell and other logicist philosophers of mathematics claimed. Poincaré thus became one of the early critics of logicism and made an influential proposal on how to steer clear of logical paradoxes (by avoiding 'non-predicative' definitions; see Poincaré 1908, 177–96). However, our focus in this chapter will remain on ideas associated with conventionalism. While Poincaré is one of its prime originators, the ideas of others have shaped the philosophical reception of conventionalism in important ways.

Duhem

One of these others was the French theoretical physicist, philosopher and historian of science Pierre Duhem (1861–1916). As a physicist, Duhem made important contributions to thermodynamics, electromagnetic theory and hydrodynamics (cf. Miller 1971 and Jaki 1984, 259–317). His wide recognition, however, rests more on his lasting influence as a historian and philosopher of science.

Poincaré in turn was an important influence for Duhem. (He also sat on the panel that accepted Duhem's doctoral dissertation in physics at the École Normale Supérieure in Paris in 1888.) In his main philosophical work, *The Aim and Structure of Physical Theory* (1906), Duhem develops views on the development of scientific theories and their relation to observation and experiment that have greatly generalized and advanced the case for the methodological claim that scientists can choose to uphold certain elements of theory even in the face of recalcitrant empirical evidence. A central and well-known part of Duhem's argument for this is his critique of 'crucial experiments' – experiments devised to bring about definite decisions between competing hypotheses. If different hypotheses give rise to different predictions about the behaviour of a certain experimental set-up, one need only perform that crucial experiment in order to determine which of the hypotheses is in error, or so traditional methodology maintained. Duhem begged to differ and claimed that a '"crucial experiment" is impossible in physics' (1906, 188). This is so because in physics, a prediction about the outcome of an experiment is never derived from one theoretical hypothesis alone. In addition to the hypothesis tested, one relies on a whole set of theoretical assumptions about the objects under study in the experiment and about the functioning of the experimental apparatus in order to predict a particular result. If the experiment is then performed and the result is *not* obtained, there is no way of knowing for sure that the fault is with the contested hypothesis rather than with one of the other assumptions. 'The only thing the experiment teaches us is that among the propositions used to predict the phenomenon and to establish whether it would be produced, there is at least one error; but where this error lies is just what it does not tell us' (ibid., 185).

To illustrate his point, Duhem discusses the example of Léon Foucault's experimental comparison in 1850 between the velocities of light in air and in water, which was widely regarded as a crucial experiment in favour of the wave theory of light and against Newton's corpuscular theory of light. While the wave theory explains the refraction of light rays at a water surface by the slower propagation of light waves in water, the corpuscular theory's explanation of the same fact involves the claim that the tiny light projectiles are accelerated by the attraction of the water particles. When Foucault determined the speed of light to be lower in water than in air, he and many of his scientific contemporaries took this to be the final blow for the hypothesis that light consists of tiny particles. Duhem objects that the mistake might just as well be hidden somewhere in the details of the Newtonian explanation of refraction, such that the corpuscular theory is erroneously attached to the idea of a higher velocity of light in water:

[T]he experiment does not tell us where the error lies. Is it in the fundamental hypothesis that light consists in projectiles thrown out with great speed by luminous bodies? Is it in some other assumption concerning the actions experienced by light corpuscules due to the media through which they move? We know nothing about that. (Duhem 1906, 187)

Duhem's general conclusion is that a hypothesis in physics is never tested against experimental evidence in isolation. Instead, it is always a larger set of physical assumptions within which the hypothesis is embedded that is subjected to empirical testing as a whole. This conclusion is nowadays sometimes called the Duhem thesis.[5] Note the similarities of Duhem's line of reasoning with Poincaré's above-quoted explanation of how scientists may, if they so choose, uphold Newton's law of gravitation even given astronomical discoveries that appear to contradict it. Duhem himself remarks that his observations explain how specific theoretical elements 'which the physicists of a certain epoch agree in accepting without test and which they regard as beyond dispute' can be saved from refutation by always choosing to modify one of the other assumptions whenever the data fail to match the predicted result (Duhem 1906, 211).[6] Duhem's thesis also entails that in physics, no experiment can ever be regarded as a definite falsification of any particular isolated theoretical conjecture. (It thereby poses a problem for Karl Popper's falsificationist philosophy of science – see Chapter 5 – and is often invoked in its criticism.)

In Duhem's own view, his arguments are about theory and experiment in physics specifically. To him, the special feature of physics that gives rise to all his methodological observations is the high degree to which the elements of physical theory are interconnected. This is mainly so because checking the observational consequences of a physical theory always requires the theoretical interpretation of an observation or a phenomenon, and this interpretation always involves other parts of physical theory (Duhem 1906, 144–6). According to Duhem, the same cannot be said about all sciences. Later generalizations of the Duhem thesis thus carry it beyond its originally intended scope (Ariew 1984).

It should be emphasized that Duhem did not single out any particular elements of physical theory and identify them as conventions. He did however highlight the freedom that physicists enjoy when they reshape theory to fit experimental results. In deciding which elements of theory to retain and which to modify in reaction to new evidence, they have to exercise their 'good sense' (Duhem 1906, 216–18). This good sense thus plays an indispensible and fundamental role for 'the entire edifice of scientific truth', according to Duhem (1903, 95).

The theory that the physicists thus establish is 'a system of mathematical propositions, deduced from a small number of principles, which aim to represent as simply, as completely, and as exactly as possible a set of experimental laws' (Duhem 1906, 19). For Duhem, physics occupies itself with regularities that are accessible by experiment. It is *not* concerned with revealing natures and entities that are ultimately responsible for these regularities (ibid., 20–1, 7–18). Instead, Duhem embraces the idea of the Austrian physicist and philosopher Ernst Mach that physical theory is first and foremost an 'economy of thought', its goal being to organize and structure empirical facts in order to facilitate their parsimonious representation (Mach 1882; Duhem 1906, 21–3). To claim for physics an aim more ambitious than mere representation and classification, such as giving explanations based on 'the nature of the elements which constitute material reality', would mean to want something that 'transcends the methods used by physics'. Physics uses 'the experimental method, which is acquainted only with sensible appearances and can discover nothing beyond them' (Duhem 1906, 10).[7] Note that this view of physical theory and Duhem's methodological observations with regard to the Duhem thesis are not unrelated. If the confrontation of theory with problematic evidence always admits more than one rational response, and the development of theory is thus 'underdetermined' by the empirical data (a term not introduced by Duhem, but by later methodologists), it seems fitting to regard the dependable content of physical theories as limited to the empirical generalizations they contain.

However, it would be a mistake to categorize Duhem's philosophy as an unambiguously anti-realist one. While he denied that the methods of physics were adequate for identifying the real constituents of nature, he affirmed their suitability for arriving at ever truer classifications of the regularities discovered by experiment. He accepted that there *are* 'hidden realities' underneath the observable data (albeit unknowable through the methods of science) and that over time the classifications of experimental laws contained in physical theory will converge to a 'natural classification' which reflects the 'real affinities among the things themselves' (Duhem 1906, 24–7, 297–8). In the light of this aspect of his philosophy, the same has been described as the attempt to find a middle way between conventionalism and scientific realism (McMullin 1990).

Nonetheless, Duhem's most influential contributions to philosophy of science remain his arguments regarding the impossibility of crucial experiments and conclusive tests of individual hypotheses. They have been taken by many philosophers of science to strengthen the case for the indispensability of non-empirical considerations in scientists' efforts to advance scientific theories.

Over and above his philosophical impact, Duhem was an outstanding historian of science. He unearthed whole schools of medieval science from the obscurity of forgotten manuscripts and thereby demonstrated the existence of a continuous tradition that led to the discoveries of Leonardo, Galileo and others. This work was instrumental in debunking the myth of the 'dark middle ages'. Throughout his life, Duhem was a devout Catholic. This fact has been used to elucidate his philosophical efforts on the one hand, as an attempt to separate physics from metaphysics, thus to secure a legitimate domain each for science and religion (cf. Deltete 2008), and his historical work on the other, as the endeavour to uncover the constructive contributions of Christian faith to the development of modern science (cf. Martin 1991).

Reichenbach

When thinkers like Helmholtz and Poincaré were reflecting on the possibility of a non-Euclidean geometry for physical space, their considerations still had the character of mere thought experiments and did not seriously challenge the use of Euclidean geometry in physics. This situation changed dramatically in 1915, with the advent of Einstein's general theory of relativity (GTR). According to GTR, the geometry of four-dimensional space-time is non-Euclidean. Physical space is described as possessing variable curvature,[8] the curvature depending on the distribution of masses and fields. The theory transforms the action of gravity into a feature of space-time geometry. Freely falling bodies are simply describing geodesics (shortest paths) in a four-dimensional space-time that is warped by the masses distributed within it, such that GTR has no more need for a gravitational force to account for those phenomena that were explained by it in classical physics. Gravitation has been 'geometrized'. For the conventionalist tradition, this development is both a triumph and a backlash. On the one hand, the conventionalist opposition to geometric apriorism is forcefully vindicated. What is more, important steps in Einstein's development of the relativity theories can be described as the discovery of conventional elements in physical theory, as we shall see. But on the other hand, the fact that geometry is now so intricately interwoven with physics seems to suggest that GTR reveals how the geometrical structure of space-time is really an empirical matter. Einstein, for one, supported geometric empiricism rather than conventionalism (Einstein 1921).

One of the first and most influential philosophers to reflect on questions concerning the conventionalist tradition in the context of relativity theory was Hans Reichenbach (1891–1953). Reichenbach had studied physics, mathematics and philosophy and in 1918/19 attended Einstein's first lecture course on GTR in Berlin. Originally strongly influenced by the neo-Kantianism then prevalent in German philosophy, he felt forced to give up apriorism in the face of GTR.

In his early book *The Theory of Relativity and A Priori Knowledge* (1920), Reichenbach still thought that Kant's idea of synthetic a priori principles shaping all our cognitive activities could be partially salvaged. He distinguished between two meanings of the Kantian a priori: 'First, it means "necessarily true" or "true for all times", and secondly, "constituting the concept of object"' (Reichenbach 1920, 48). Under the impression of the revolutionary changes in geometry that were brought about by GTR, he considered the a priori in the first sense irretrievably lost, but in the second sense worth retaining. The constitutive principles that Reichenbach described, however, were not constitutive for human experience as such, as Kant's had been, but constitutive for the kind of scientific knowledge attainable within the framework of a particular theory. In particular, he identified these constitutive elements in the 'coordinative principles' that establish the connection between the mathematical structures used by a theory (such as a certain geometry) with their empirical content – for example, by establishing how the length of an object is to be determined through a certain interpretation of measuring procedures (ibid., 40). These ideas have since been interpreted as an attempt to recover the conception of a 'relativized a priori' (Friedman 2001, 30–1, 72). Reichenbach himself soon became convinced that the direction of his thoughts had little to do with Kantianism any more.[9] His highly influential book of 1928, *The Philosophy of Space and Time*, is stripped of the Kantian allusions but still emphasizes that all scientific knowledge has to be informed by both empirical evidence and constitutive principles, now called coordinative definitions.

One example that Reichenbach employs to explain the need for coordinative definitions is exactly the one that Poincaré uses in his arguments for geometric conventionalism. To relate the metric of the mathematical structures used in our physical theory of space to empirical reality, we have to define what kinds of bodies will count as measuring rods under which circumstances (in particular, under which circumstances we will assume them to retain their length). This definition is based on empirical observations, giving us confidence that we can consistently use certain bodies in this way. For example, we observe that two rods can be brought to match at different places, independently of the paths that are taken to move them around. However, '[w]hen we add to this empirical fact the definition that the rods shall be called equal in length when they are at *different places*, we do not make an inference from the observed fact; the addition constitutes an independent convention' (Reichenbach 1928, 16–17).

According to Reichenbach, Einstein's breakthrough consisted in part in 'the discovery of the definitional character of the metric in all its details', that is, in the realization that coordinative definitions are required also for the comparison of lengths in systems of different states of motion and for the comparison of time intervals occurring at different locations in space (ibid., 177, cf. 15).[10] He calls this the 'epistemological foundation' of the theory of relativity. The physical part of the theory that completes this is 'the hypothesis that natural measuring instruments follow coordinative definitions different from those assumed in the classical theory' (ibid., 177).

Does this mean that Einstein's choice of a non-Euclidean geometry for GTR is based on completely arbitrary conventions? Reichenbach does not think so. He develops a principled account for the choice of a geometry that lets Einstein's choice emerge as a special case. The account is based on the distinction between universal and differential forces (ibid., 13–14, 22–8). Universal forces affect all bodies in the same way and they cannot be shielded against. All other forces are called differential. The effect of temperature on solid bodies in our actual world can be regarded as a differential force in this sense, because it acts differently on a copper rod and an iron rod. In contrast, the effects of temperature in the Sphereland thought experiment introduced in section 2 provide a good example for a universal force. Sphereland also illustrates the fact that in the presence of a truly universal force, physical theory can be reformulated in such a way that the universal force and its effects are eliminated by choosing a suitable geometry. (We had assumed that the Spherelanders' own science would take that course by adopting hyperbolic geometry.) Reichenbach stipulates that this is exactly the course that *should* be taken when a geometry is chosen. He demands that the crucial coordinative definition that lays down what counts as a rigid rod in empirical reality may include correction factors for differential forces, but not for universal ones (ibid., 22). The geometry that must be selected as a result is the one that in effect sets all universal forces to zero. The Spherelanders are thus committed to the hyperbolic option, and we to the variable curvature geometry of GTR. As a result of Einstein's equivalence principle, gravity in GTR must affect all substances in the same way. Its effect on the length of measuring rods should therefore be set to zero. By thus adopting the suitable geometry, we spare ourselves the talk of measuring rods contracting in gravitational fields. 'This universal effect of gravitation on all kinds of measuring instruments defines therefore a single geometry' (ibid., 256).

Reichenbach, while emphasizing the inevitable contribution of conventional principles to every physical theory, attached great importance to the fact that the conventional choices should not be regarded as arbitrary. He therefore rejected the term 'conventionalism' for his view (Reichenbach 1922, 38) and discreetly distanced himself from Poincaré, who he thought had overemphasized arbitrariness (ibid.; Reichenbach 1928, 36–7).

Reichenbach's importance for modern philosophy of science far exceeds the contributions in the focus of this chapter. Though he was never himself a member of the Vienna Circle, he was one of the key thinkers of logical empiricism (see chs 3 and 4). A distinctive feature of several of his philosophical works is an emphasis on the concept of probability. For example, he advocated a probabilistic version of the early logical positivists' verifiability criterion of meaning. According to Reichenbach's version, a proposition has meaning if it is possible to determine a degree of probability for it, and two sentences have the same meaning if they obtain the same degree of probability by every possible observation (Reichenbach 1938, 54). This theory was intended

to avoid the problem that demanding verifiability of meaningful propositions would render the general statements of scientific theories meaningless, because they can never be verified in a strict sense. Reichenbach
made pioneering contributions to several specialties that continue to
engage philosophers of science, including the philosophical foundations of probability, causation, the direction of time and the philosophy
of quantum mechanics. He emigrated from Germany after the rise to
power of the Nazi Party in 1933. After five years at Istanbul University,
he became full professor at the University of California at Los Angeles,
where he remained for the rest of his life, and thus he also played a role
in establishing philosophy of science in North America.

The conventionalist legacy

As regards geometric conventionalism, the endeavour to establish an
interpretation of GTR that stresses the role of conventions was taken
up by Adolf Grünbaum (1973) and others (see Juhl and Loomis 2006 for
an overview). While one might say that the mainstream of philosophy
of physics today sides with geometric empiricism, the debate should
not be regarded as settled. Recently, Yemima Ben Menahem (2006,
80–136) argued that geometric conventionalism remains a live option.
Among other things, she appeals to the fact that Richard Feynman
(1971) and other physicists have offered expositions which first develop
the equations of GTR from considerations about gravitational fields;
the identification of the Einstein tensor as determining the geometrical
structure of space-time is added only at a later stage, as an interpretation of the equations.

Ben Menahem also proposes to distinguish two general elements
that characterize the philosophical legacy of conventionalism in a wider
sense: On the one hand, the conventionalist thesis has been extrap-
olated from its earliest articulations to a general explanation of how
mathematical and logical truths are grounded in convention. The pre-
dominant author here is Carnap (see Chapter 3). On the other hand, the
conventionalists' core idea of underdetermination of theory by evidence
has been widely received and developed further, in particular, by Quine.
Ironically, the second of these two conventionalism-based traditions
ultimately turned into a trenchant critique of the first: If the Duhem

thesis is radicalized into the claim that we can choose *any* part of the theory and hold it true come what may, then the project of sharply distinguishing the logico-mathematical framework of a theory which is due to convention from the theory's content which is due to experience seems to become a futile undertaking.

Further reading

Ben Menahem (2006) gives a wide-ranging and detailed study of conventionalist philosophies. The development of non-Euclidean geometries and their interpretations that led up to Poincaré's insights are presented in Torretti (1978). Poincaré's philosophy is explored in Zahar (2001) and Folina (1992). On Duhem, see Martin (1991), Jaki (1984) and the papers collected in Ariew and Barker (1990). Ryckman (2007) surveys Reichenbach's (and other logical empiricists') contributions to philosophy of physics. On Reichenbach, see also the essays in Salmon and Wolters (1994), Spohn (1991) and Salmon (1979).

Notes

1 The name is borrowed, not from Helmholtz, but from Edwin Abbott's (1884) literary fiction about a two-dimensional world.
2 A spherical surface is an *exact* illustration of elliptical geometry only if we identify every point on the surface with its antipode. That is because an axiom of all (Euclidean and non-Euclidean) geometries demands that there be exactly one straight line connecting two distinct points. If a point and its antipode are considered distinct, this axiom is violated by spherical geometry, because there are an infinite number of great circles connecting the two.
3 On Poincaré's mathematical and scientific accomplishments, see Dieudonné 1975.
4 A further restriction that bears mentioning derives from the fact that Poincaré was a steadfast adherent of Sophus Lie's group theoretical classification of geometries, as exposed, for example, in Poincaré 1898 (see also Poncaré 1902, 46–7). Lie and others had characterized geometries by the specific group of displacements of a rigid body allowed in each of them. Poincaré thought that Lie's groups describe an idea of geometry that 'pre-existed' in our minds (1902, 87). This in effect limits the choice to geometries of constant curvature (see fn 8, below).

5　It was later radicalized by Willard Van Orman Quine, who expanded the body of knowledge that is put to test in every empirical check (not just in physics) to include even our mathematical and logical beliefs. This amplified version is also called the Duhem-Quine thesis.

6　Another sign of Poincaré's influence is Duhem's other example for an alleged crucial experiment, Otto Wiener's 1891 experiment on the plane of oscillation in polarized light (Duhem 1906, 184–6). Poincaré had criticized its purported role as a crucial experiment in a paper of 1891, and Duhem had taken up Poincaré's critique and generalized it as early as 1892 (Duhem 1892; cf. Martin 1991, 104–6, and Ben Menahem 2006, 72–3).

7　Apart from the view that physics aims at revealing the ultimate constituents of nature and giving explanations in terms of them, there is a second view of physical theories that Duhem was eager to criticize. It is the view that physics aims at constructing mechanical models that render physical phenomena understandable to the human mind, which Duhem considered to be manifest in the work of many of his English contemporaries (Duhem 1906, 69–75). While the English physicists seemed unperturbed by the fact that their many mechanical models (for example for the electromagnetic ether) often contradicted each other, Duhem regarded this as frustrating the objective of 'the unity of science', which every physicist 'naturally aspires to' (ibid., 103).

8　Because two-dimensional non-Euclidean geometries can be modelled and illustrated with the aid of curved surfaces (see the Flatland thought experiment above), 'curvature' has come to be used as a technical term for a central mathematical characteristic of geometries. The curvature of elliptical geometries (where the angle sum of a triangle is larger than 180 degrees) is positive, the curvature of hyperbolic geometries (with angle sums smaller than 180 degrees) negative. Euclidean geometry has zero curvature. The possibility of geometries with a curvature that varies across space was already described by Riemann in 1854.

9　It seems to have been Moritz Schlick who convinced Reichenbach of this by correspondence; see Gerner 1997, 53–5.

10　The question whether the metric, and in particular the definition of simultaneity in the special theory of relativity, should be regarded as conventional or not has led to a controversy that lasts to this day; see Janis 2008 for overview.

References

Abbott, Edwin (1884), *Flatland: A Romance of Many Dimensions*. 6th edn. New York: Dover 1977.

Ariew, Roger (1984), 'The Duhem Thesis'. *British Journal for the Philosophy of Science*, 35: 313–25.

Ariew, Roger and Peter Barker (eds) (1990), *Pierre Duhem: Historian and Philosopher of Science*, special issues of *Synthese*, 83 (2–3).

Ben Menahem, Yemima (2006), *Conventionalism*. New York: Cambridge University Press.

Deltete, Robert (2008), 'Man of Science, Man of Faith: Pierre Duhem's "Physique de Croyant"'. *Zygon*, 43: 627–37.

Dieudonné, Jean (1975), 'Poincaré, Jules Henri', in *Dictionary of Scientific Biography*, ed. Charles C. Gillispie, vol. 11. New York: Scribner, pp. 51–61.

Duhem, Pierre (1892), "Some Reflections on the Subject of Physical Theories", in *Essays in the History and Philosophy of Science*, ed. and trans. Roger Ariew and Peter Barker. Indianapolis: Hackett 1996, pp. 1–28.

— (1903), *The Evolution of Mechanics*, trans. Michael Cole. Alphen aan den Rijn: Sijthoff and Noordhoff, 1980.

— (1906): *The Aim and Structure of Physical Theory*, trans. Philip P. Wiener. New York: Atheneum 1962.

Einstein, Albert (1921), 'Geometry and Experience', trans. S. Bargmann, in *Ideas and Opinions*. New York: Crown 1945, pp. 232–45.

Ewald, William (ed.) (1996), *From Kant to Hilbert: A Source Book in the Foundations of Mathematics*. 2 vols. Oxford: Clarendon.

Feynman, Richard (1971), *Lectures on Gravitation*. Pasadena: California Institute of Technology.

Folina, Janet (1992), *Poincaré and the Philosophy of Mathematics*. Houndmills: MacMillan.

Friedman, Michael (2001), *The Dynamics of Reason*. Stanford, CA: CSLI Publications.

Gerner, Karin (1997), *Hans Reichenbach - sein Leben und Wirken, eine wissenschaftliche Biographie*. Osnabrück: Phoebe.

Grünbaum, Adolf (1973), *Philosophical Problems of Space and Time*. 2nd edn. Boston: Reidel.

Helmholtz, Hermann von (1876), 'The Origin and Meaning of Geometrical Axioms', in Ewald (1996), vol. 2, pp. 663–89.

Jaki, Stanley L. (1984), *Uneasy Genius: The Life and Work of Pierre Duhem*. The Hague: Nijhoff.

Janis, Allen (2008), 'Conventionality of Simultaneity', *Stanford Encyclopedia of Philosophy* (Fall 2008 edn), Edward N. Zalta (ed.), http://plato.stanford.edu/archives/fall2008/entries/spacetime-convensimul/.

Juhl, Cory and Eric Loomis (2006), 'Conventionalism', in *The Philosophy of Science: An Encyclopedia*, ed. Sahotra Sarkar and Jessica Pfeifer, vol. 1. New York: Routledge, pp. 168–75.

Kant, Immanuel (1781/87), *Critique of Pure Reason*, ed. and trans. P. Gruyer and A. W. Wood. Cambridge: Cambridge University Press, 1998.

Lewis, David K. (1969), *Convention: A Philosophical Study*. Cambridge, MA: Harvard University Press.

Mach, Ernst (1882), 'The Economical Nature of Physical Inquiry', in *Popular Scientific Lectures*, trans. T. J. McCormack. LaSalle, IL: Open Court 1986, pp. 186–214.

Martin, R. and D. Niall (1991), *Pierre Duhem: Philosophy and History in the Work of a Believing Physicist*. LaSalle, IL: Open Court.

McMullin, Ernan (1990), 'Comment: Duhem's Middle Way', *Synthese*, 83: 421–30.

Miller, Donald G. (1971), 'Duhem, Pierre-Maurice-Marie', in: *Dictionary of Scientific Biography*, ed. Charles C. Gillispie, vol. 4. New York: Scribner, pp. 225–33.

Poincaré, Jules Henri (1898), "On the Foundations of Geometry", in: Ewald (1996), vol. 2, pp. 982–1011.

— (1902), *Science and Hypothesis*, trans. W. J. Greenstreet. New York: Dover, 1952.

— (1905), *The Value of Science*, trans. Bruce Halsted. New York: Dove, 1958.

— (1908), *Science and Method*, trans. Francis Maitland. New York: Dover 1952.

Reichenbach, Hans (1920), *The Theory of Relativity and A Priori Knowledge*, trans. Maria Reichenbach. Los Angeles: University of California Press, 1965.

— (1922), 'The Present State of the Discussion on Relativity: A Critical Investigation', in *Modern Philosophy of Science: Selected Essays*, trans. and ed. Maria Reichenbach. London: Routledge 1959, 1–45.

— (1928), *The Philosophy of Space and Time*, trans. Maria Reichenbach and John Freund. New York: Dover, 1958.

— (1938), *Experience and Prediction: An Analysis of the Foundations and the Structure of Knowledge*. Chicago: University of Chicago Press.

Ryckman, Thomas (2007), 'Logical Empiricism and the Philosophy of Physics', in *The Cambridge Companion to Logical Empiricism*, ed. Alan Richardson and Thomas Uebel. Cambridge: Cambridge University Press, pp. 193–227.

Salmon, Wesley C. (ed.) (1979), *Hans Reichenbach: Logical Empiricist*. Dordrecht: Reidel.

Salmon, Wesley C. and Gereon Wolters (eds) (1994), *Logic, Language, and the Structure of Scientific Theories: Proceedings of the Carnap-Reichenbach Centennial, University of Konstanz, 21–24 May 1991*. Pittsburgh: University of Pittsburgh Press.

Spohn, Wolfgang (ed.) (1991), *Erkenntnis Orientated: A Centennial Volume for Rudolf Carnap and Hans Reichenbach*, special issues of *Erkenntnis*, 35 (1–3).

Torretti, Roberto (1978), *Philosophy of Geometry from Riemann to Poincaré*. Dordrecht: Reidel.

Zahar, Elie (2001), *Poincaré's Philosophy: From Conventionalism to Phenomenology*. Chicago: Open Court.

CHAPTER 3

THE VIENNA CIRCLE: MORITZ SCHLICK, OTTO NEURATH AND RUDOLF CARNAP

Friedrich Stadler

The Vienna Circle (as the so-called Schlick Circle came to be known) consisted of a group of about three dozen thinkers coming from the natural and social sciences, logic and mathematics, which met regularly in Vienna between the two world wars to discuss philosophy and its relation to the sciences. The work of this group constitutes one of the most important and most influential philosophical contributions of the twentieth century, in particular in the development of analytic philosophy and history and philosophy of science (Stadler 2001, 2003a; Richardson and Uebel 2007).

Prior to World War I, the predecessor of the later Vienna Circle had begun to take shape both as an organization and as a philosophy (Uebel 2000). Within a discussion circle (including Frank, Hahn and Neurath inter alia) at a coffeehouse, traditional 'academic philosophy' grew more scientific. This so-called First Vienna Circle met regularly as of 1907 to discuss the synthesis of empiricism and symbolic logic as modelled after Mach, Boltzmann and the French conventionalists Pierre Duhem and Henri Poincaré (see Chapter 2).

This early phase in the development of logical empiricism can also be interpreted as an anti-Cartesian turn in epistemology and philosophy of science, which undermined both the synthetic a priori and the secure foundations of knowledge. In the middle of the permanent crisis of philosophy between reform and revolution in society and science, the further development of this 'scientific philosophy' had, in any case, been initiated.

The Vienna Circle was first publicly announced in 1929 with the publication of what came to be called its manifesto, *Wissenschaftliche Weltauffassung. Der Wiener Kreis* (The Scientific Conception of the World: The Vienna Circle), edited by the Verein Ernst Mach (Ernst Mach Society) and authored by Rudolf Carnap, Hans Hahn and Otto Neurath (Carnap et al. 1929). This group was essentially a modernist movement, at the centre of which was the so-called Schlick Circle, a discussion group organized in 1924 by Moritz Schlick. Rudolf Carnap, Herbert Feigl, Philipp Frank, Kurt Gödel, Hans Hahn, Otto Neurath, Felix Kaufmann, Viktor Kraft, Karl Menger, Friedrich Waismann and Edgar Zilsel belonged to its inner circle. Their meetings were also attended by Olga Taussky-Todd, Olga Hahn-Neurath, Rose Rand, Gustav Bergmann and Richard von Mises, and on several occasions by visitors such as Hans Reichenbach, Alfred J. Ayer, Ernest Nagel, Willard Van Orman Quine and Alfred Tarski. Members of the periphery, most of them as participants and occasional guests, were Egon Brunswik, Karl Bühler, Josef Frank, Else Frenkel-Brunswik, Heinrich Gomperz, Carl Gustav Hempel, Eino Kaila, Hans Kelsen, Charles Morris, Arne Naess, Karl Popper, Frank P. Ramsey, Kurt Reidemeister and the alleged 'genius', Ludwig Wittgenstein, who was not attending the Schlick Circle but met only Schlick and Waismann regularly and had a special influence on some members of the group. In addition, the mathematician Karl Menger organized in the years 1926–36 an international 'Mathematical Colloquium', which was attended by Kurt Gödel, John von Neumann and Alfred Tarski, among many others (Menger 1994).

This international and interdisciplinary discussion circle was pluralistic and committed to the ideals of the Enlightenment. It was unified by the aim of making philosophy scientific with the help of modern logic on the basis of experimental and everyday experience. The general aims of the movement were expressed in its publications, such as the two book series – Schriften zur Wissenschaftlichen Weltauffassung (Publications on the Scientific Conception of the World), 1929–37, 11 volumes; and Einheitswissenschaft (Unified Science), 1933–38, 7 volumes – the journal *Erkenntnis*, 1930–40 (the 1939 volume was called *Journal for Unified Science*); and the *International Encyclopedia of Unified Science*, 1938–70 (edited by Neurath, Carnap and Morris, 1971).

Given this story of scholarly success, the fate of the Vienna Circle was tragic. The Verein Ernst Mach was suspended in 1934 by Austro-Fascism for political reasons; Schlick was murdered in 1936 by a fanatic student, and around this time, many members of the circle were forced to leave Austria for racial and political reasons. Thus, soon after Schlick's death, the circle disintegrated. As a result of the emigration of so many of its members and adherents, however, the circle's ideas became more and more widely known, especially in northern Europe, Britain and North America, where they contributed hugely to the emergence of modern philosophy of science (Timms and Hughes 2003; Hardcastle and Richardson 2003; Manninen and Stadler 2010). In Germany and Austria, however, the break that was caused by the forced emigration of the Vienna Circle's members was felt on the philosophical and mathematical scene for a long time (Heidelberger and Stadler 2003; Stadler 2010).

Historically, Mach's philosophy provided the foundation for the development of the positions adopted within the Vienna Circle. The expression 'logic of science' (Carnap's 'Wissenschaftslogik'; 1934a), known since the mid-1930s as 'philosophy of science', was later used to describe these positions. This implied a general scientific conception of philosophy as well as an attempt to provide a philosophy for all sciences (including the humanities). In addition, within the Vienna Circle, philosophy was regarded both as a form of linguistic analysis and as a discipline drawing on the foundations of the natural and social sciences.

At the same time there were divergences of philosophical approaches within the Vienna Circle. Schlick and others defended a methodological dualism of philosophy and science, and a group that included Carnap and Neurath sought to integrate philosophy altogether within a scientific conception of the world or encyclopedic unity of the sciences. In Schlick's view, the classical philosophical positions of empiricism and rationalism were integrated with the help of modern logic and mathematics, but a distinction between philosophy and science still remained. Neurath's more radical physicalism or 'encyclopedism' of logical empiricism aimed at overcoming philosophy itself within his collective project of an international encyclopedia of the unity of science (Neurath 1946a). This divergence in philosophical approaches left room for debates within the circle on such topics as the merits of phenomenalist and physicalist languages, coherence and correspondence theories of truth, logical syntax and semantics, verification and confirmation and ideal and natural languages. At the same time there was a certain consensus on the merits of logical analysis of language, a fallibilist epistemology, a scientific attitude to the world and the unity of scientific explanation and knowledge in general.

The rivalry between Schlick's 'consistent empiricism' and Neurath's physicalist (later empiricist) unified science is a complex matter. Certain views were held by both, such as the view of philosophy as a critique of language in accordance with Wittgenstein's philosophy of the *Tractatus* (1922). However, while the principles of verification, logical atomism and the picture theory of language are constitutive features of the entire movement, by themselves they do not characterize the Vienna Circle. Theoretical elements like logicism, verifiability, methodological phenomenalism and physicalism, a fallibilist theory of knowledge, conventionalism and realism, together with an empiricist encyclopedism, were cornerstones of the internal pluralistic development of logical empiricism from the 1930s onwards. This development also reflected the influence of Neurath's historico-pragmatic point of view within the circle. In particular, the objection towards any dualism of 'language' and 'world' (as *Wirklichkeitsphilosophie*), with the attendant denial of any absolute 'foundation of knowledge' (Schlick 1934), is representative of this non-reductive naturalism and methodological holism in the spirit of Pierre Duhem's and Henri Poincaré's philosophy of science. This form of relativism and naturalism already anticipated the historical turn after World War II in the philosophy of science (cf. Kuhn 1962), which contributed to overcoming the linguistic turn and the so-called received view of philosophy of science.

Nevertheless, the rejection of synthetic a priori judgments remained an important element of the Vienna Circle. According to Russell and Whitehead in the *Principia Mathematica*, symbolic logic and mathematics were regarded as purely analytic and a priori (independent of any experience). Analytic truths of these kinds were contrasted with contingent statements of the natural sciences and ordinary everyday experience, as synthetic a posteriori judgments. But there was no further class of synthetic a priori judgments; instead there was thought to be an important class of 'meaningless' sentences, sentences without any cognitive content. The elements of this class were seen as 'metaphysical', in the sense that they are not part of knowledge at all even though they may express some realm of commonsense experience (for a moderate summary, cf. von Mises 1951).

This position of the classical Vienna Circle is most prominently represented by Carnap's 'Elimination of Metaphysics Through Logical Analysis of Language' (1931), which developed a program for a unified rational reconstruction of science. But the question as to whether an empirical basis could serve as the foundation for all knowledge received strongly divergent answers from coherence theorists about truth influenced by Neurath and correspondence theorists influenced by Schlick (Hempel 1993). Also, the apparently strict distinction between analytic and synthetic sentences was questioned (Menger 1979, 1–60). The ideal of one language of science, logic, and mathematics was radically weakened within the Vienna Circle itself with Menger's and Carnap's principle of tolerance long before Quine (1953) put forward his critique, 'Two Dogmas of Empiricism'. Thus, contrary to popular belief, a heterogeneous pluralism of views was in fact characteristic of the Vienna Circle; for example, regarding ethics (Schlick, Menger, Kraft), the alternatives of realism versus positivism (Schlick, Carnap, Feigl, Kraft, Kaufmann), verificationism versus falsificationism (both positions criticized by Neurath, especially against Popper's *Logic of Scientific Discovery*) and, last but not least, matters of ideological and political preference – for example, conservative liberalism versus leftist socialism. In the later period of the Vienna Circle the contested verification principle was gradually abandoned and replaced by some form of a probabilistic confirmation methodology based on the principle of 'connectibility' (von Mises 1951).

The unity of science movement, with its six International Congresses for the Unity of Science (1935–41) and the ambitious publication project *International Encyclopedia of Unified Science* (1938–70), had a broader cultural meaning and goal, most notably the attempt to improve the human condition and to promote social reform and the intellectual struggle against irrationalism and totalitarian *Weltanschauungen*. It was a manifestation of a late-Enlightenment conception of science with a socially inspired anti-metaphysics. Between the two world wars metaphysics was seen as a correlative feature of German idealism as well as of (Austro-)Fascist 'universalism', as represented by the economist Othmar Spann.

The practical impulse behind this therapeutic destruction of metaphysical systems, then, was the desire for a scientific attitude based on human experience, directed against the zeitgeist of totalitarianism and cultural pessimism (as criticized already by Neurath in 1921 and 1931). Therefore, traditional philosophy, first of all, had to be reduced to a critical analysis of language, because most proponents of logical empiricism thought that an exact and sober usage of the scientific language is a precondition for all problem-oriented philosophizing – and moreover a sort of moral obligation.

Social criticism and collective work in philosophy of science formed a programmatic unity striving for a sweeping improvement of the human condition. Whereas in the natural sciences considerable progress had already been made, the situation in the social and cultural sciences, influenced by the ongoing *Methodenstreit* since the turn of the century (Kaufmann 1936), was not so transparent. Although some members of the Vienna Circle, like Kaufmann, Neurath and Zilsel, contributed essentially to this neglected field, their publications have been largely ignored in the historiography on the circle for a long time. In this respect it is worth mentioning that, after the disintegration of the Vienna Circle, its former members still occasionally made reference to the 'scientific conception of the world' when speaking about general ideological questions. For example, Carnap spoke about 'scientific humanism' as a view shared by the majority of the logical empiricists (Carnap 1963, 81ff.).

After the dissolution of the Vienna Circle, the forced migration of most of its members and the dispersion of the logical empiricist movement from its centres in central Europe, the twin aims of a transformation of philosophy and the establishment of philosophy of science could only be envisaged once the ties to their previous cultural context and audience had been severed. But even in these difficult times the proponents of the exiled Vienna Circle organized six well-attended, prestigious international conferences, the International Congresses for the Unity of Science: Paris (1935 and 1937), Copenhagen (1936), Cambridge, UK (1938), Cambridge, Massachusetts (1939) and Chicago (1941). One can thus say that the demise of the Vienna Circle in the German-speaking world was accompanied by the transformation of Viennese *Wissenschaftslogik* into philosophy of science in the Anglo-Saxon scientific community, converging with North American currents like (neo)pragmatism.

Despite this pluralism is it still possible to find a sort of basic agreement here – one that unites the members of the Vienna Circle, both the central figures and those on the periphery? First of all, it is a way of philosophizing based on linguistic analysis and a great amount of problem-oriented, open-ended discussion. This was experienced personally by Arne Naess, who focused several times on the Vienna Circle's 'thought style' which, in (not only) his opinion, leads to an inherent 'pluralism of tenable worldviews' (Naess 2003). Secondly, the use of an unambiguous language, together with exact methods, is certainly a main legacy of the circle and those associated with it. It is only when this exact formal approach is adopted that the content and positions can be constructively criticized and refuted – a characteristic which most current modern and postmodern philosophies lack.

The explicit and hidden history of the Vienna Circle from *Wissenschaftslogik* to recent philosophies of science documents the wide range, pluralism and diversity of its heritage and message. Be it called 'scientific philosophy' (as initiated by Schlick), 'scientific humanism' (according to Carnap) or a modern encyclopedism as a 'republic of scholars' (following Neurath), it is a guide to an intellectual journey, which continues through the present day and probably on into the future.

Moritz Schlick – between nature and culture

Friedrich Albert Moritz Schlick was born on 14 April 1882, the third and youngest son of Protestant parents. He studied natural science and mathematics at universities in Heidelberg, Lausanne and Berlin. In 1904 he completed his Ph.D. under Max Planck, who regarded him as one of his favourite students, with a thesis 'Über die Reflexion des Lichtes in einer inhomogenen Schicht' ('The Reflection of Light in an Inhomogeneous Layer') in mathematical physics. After his first book, *Lebensweisheit* (Life Wisdom), appeared in 1908, Schlick devoted his efforts for two years to studying psychology in Zurich. He completed his habilitation in 1911 at the University of Rostock with the study *The Nature of Truth According to Modern Logic*. During his ten years of academic activity in Rostock, first as private lecturer and then as professor (1917), Schlick worked on the reform of traditional philosophy against the backdrop of the revolution in natural science. He became friends with Albert Einstein, whose theory of relativity he was one of the first to study philosophically.

In 1918 his major study, *Allgemeine Erkenntnislehre (General Theory of Knowledge)*, was published. In 1922, on Hans Hahn's initiative, Schlick was appointed to Vienna as the successor of Boltzmann and Mach, to the chair for natural philosophy (philosophy of the inductive sciences). From 1924 on, Schlick organized, at the suggestion of his students Herbert Feigl and Friedrich Waismann, a regular discussion circle, later known as the Vienna Circle, which first met privately, then in the Department of Mathematics in Vienna. In addition to extensive research and teaching activities Schlick was also active in adult education: as a member of the Ethical Society and, above all (1928–34), as chairman of the Verein Ernst Mach. In spite of numerous interventions he could not prevent its dissolution after 12 February 1934 for political reasons. From 1926 on, Schlick came in personal contact with Ludwig Wittgenstein, who influenced him in a significant way. In 1929 he refused an attractive call to Bonn (at his students' request) and then spent several months as visiting professor at Stanford, and later (1931–32) in Berkeley, California.

On June 22, 1936 Moritz Schlick was murdered on the steps of the Vienna University by a former student who acted on private and Weltanschauung motives. Schlick was at the apogee of his influential life as a scholar. The student was released before the end of his sentence by the Nazis and lived as a free citizen in Austria after 1945. This act marked the definitive demise of the Vienna Circle, whose last members were forced to emigrate after the Anschluss (1938).

Schlick accorded philosophy an important and independent function in relation to the natural and social sciences right up to his untimely passing. Independent of this, he also embodied the prototype of the liberal, cosmopolitan intellectual in the midst of a National Socialist revolution.

In the *Philosophen-Lexikon* (1950) we find an entry written by Schlick himself. It begins with the following programmatic sentence: 'Schlick attempts to justify and construct a consistent and entirely pure empiricism,' which, unlike its early forms, is reached by applying modern mathematics and logic to reality. Schlick continues:

> From there, and with the help of an analysis of the process of knowledge, the 'General Theory of Knowledge' arrives first at a clear distinction between the

rational and the empirical, the conceptual and the intuitive. Concepts are mere symbols that are attributed to the world in question; they appear in 'statements' ordered in a very particular way, by which these are able to 'express' certain structures of reality. Every statement is the expression of a fact and represents knowledge insofar as it describes a new fact with the help of old signs – in other words, with a new combination of terms which have already been used in other regards. The ordering of reality, . . . is determined solely by experience, for which reason there exists only empirical knowledge. The so-called rational truths, then, purely abstract statements such as the logical-mathematical ones . . . are nothing more than rules of signs which determine the syntax of the language (L. Wittgenstein) which we use to speak about the world. They are of purely analytic-tautological character and therefore contain no knowledge; they say nothing about reality, but it is for precisely this reason that they can be applied to any given fact in the world. Thus, knowledge is essentially a reproduction of the order, of the structure of the world; the material or content belonging to this structure cannot enter it; for the expression is, after all, not the thing itself which is being expressed. Therefore, it would be senseless to attempt to express the 'content' itself. Herein lays the condemnation of every variety of metaphysics; for it is precisely this that metaphysics has always wanted, in having as its goal the cognizing of the actual 'essence of being'. (Schlick 1950, 462f.)

This short text has its origins in Schlick's Viennese period, as indicated by his reference to Wittgenstein. But it also represents the essence of Schlick's main work on epistemology, his *General Theory of Knowledge* (Schlick 1918–25), as a manifestation of the specific sort of independent thinking with the typical duality of philosophy and the sciences – by the way, in contrast to the oft-claimed congruence with the *Tractatus* philosophy.

Thus, Schlick (1950, 463) concludes:

Philosophy is not a science, even though it pervades all sciences. Because while these latter consist of systems of true assertions and contains knowledge, philosophy consists in the search for the meaning of the statements and creates understanding, which leads to wisdom.

The quoted passages do not allow us to recognize any priority between nature or culture, nor between theoretical or moral philosophy. This is because, according to Schlick, ethics and aesthetics can be done in congruence with the concept of his 'consistent empiricism'. Hence,

it makes no sense to speak of 'absolute' values; only the evaluative behaviors actually practiced by human beings can be the object of study. Based on this

standpoint arises a new justification for a kind of eudaimonism, which moral
principle reads more or less so: increase your happiness! (ibid.)

This quote expresses a further characteristic, which is clearly different
from Wittgenstein's philosophy of the ineffable, and the abstinence
from moral-philosophic issues among some members of Vienna Circle
and Berlin Group. This characteristic is elaborated in Schlick's monographs from *Lebensweisheit* (1908) to his *Problems of Ethics* (1930).

A most decisive episode was Schlick's encounter with Albert Einstein,
with whom he had been exchanging letters on a regular basis from 1915
on. Initially, Schlick was viewed by Einstein as being *the* philosophical
interpreter of his theory of relativity. Later on, Einstein himself is seen –
alongside Russell and Wittgenstein – as a model and a proponent of
the *Wissenschaftliche Weltauffassung* (1929). particularly with regard to
the conclusion he drew in his *Geometrie und Erfahrung* (Geometry and
Experience): 'Insofar as the statements of mathematics refer to reality,
they are not certain, and insofar as they are certain, they do not refer to
reality' (Einstein 1921, 119f.).

Schlick ended up finding himself in disagreement with his former
idol after having turned to Wittgenstein's philosophy of language from
1924 on – before he had produced one of the most important presentations of Einstein's writings in *Space and Time in Contemporary Physics*
(1917), which went through four printings. Their later differences – from
the mid-1920s onward – had to do with central positions from realism
(Schlick 1932–33) to conventionalism and to the principle of causality
in the context of a philosophy which, via linguistic analysis, views the
problem of reality as a genuine pseudo-problem in the sense of Rudolf
Carnap (1928) and Otto Neurath. After Schlick's Rostock period (up to
1921) and Kiel (1921), the then 40-year-old philosopher was appointed
to the chair for natural philosophy (philosophy of the inductive sciences)
in Vienna in the tradition of Mach and Boltzmann in 1922. The mathematician Hans Hahn, who was the teacher of Karl Menger and Kurt
Gödel, was mainly responsible for this innovative step. This move also
represented an attempt to give an institutional platform and provide
an intellectual figure for the further development of this scientific philosophy since the first Vienna Circle (with Frank, Hahn, and Neurath). In
this period, Schlick also read Wittgenstein's *Tractatus* in the seminars
held by the mathematicians Hans Hahn and Hans Reidemeister and also

addressed Russell's philosophy of logical atomism and neutral monism (McGuinness 1985).

By 1924 all the intellectual and institutional foundations for the formation of the Schlick Circle had been laid. The writings of Frege, Russell and Whitehead, and Wittgenstein constituted the theoretical frame of reference against the backdrop of the synthesis of Mach and Boltzmann on the one hand and Duhem and Poincaré on the other. This was a process that was most lucidly described by Philipp Frank, Einstein's successor in Prague (Frank 1949).

It was in the air when in fall 1924 Schlick, at the instigation of his talented students Herbert Feigl and Friedrich Waismann, decided to introduce a regular discussion forum ('evening circle') at the University of Vienna. This forum remained in existence up until his death and later went down in the history of philosophy and science as the Vienna Circle (Stadler 2001, 2003). This institutionalization of the Schlick Circle between 1924 and 1929 was characterized by the discussion and encounter with Wittgenstein's early philosophy and finally with Rudolf Carnap's *Logical Structure of the World* (1928) – which was inspired by neo-Kantianism, Gestalt theory and set theory, based on Mach and Russell.

Until the second edition of his *Allgemeine Erkenntnislehre* (1925), Schlick was indebted to a form of critical realism. But in the 1930s he came to embody, together with Waismann, that wing of the circle inspired but not determined by Wittgenstein. The distinct notions of philosophy and concepts of ethics and aesthetics remained characteristic features of Schlick's philosophizing; he conceived of philosophy as a system for expressing the most general principles inherent in the sciences. Philosophy was seen as a clearing activity, involving the logical analysis of statements within the individual sciences, for the sake of both overcoming metaphysics and clarifying meaning.

We can thus schematically trace out a movement in Schlick's work from his initial, realistic position, via the 'linguistic turn' with verificationism, to the more liberal, still correspondence-theoretical position he held to till the end of his life. This was also a concrete distancing from Neurath's 'non-philosophical' and coherence-theoretical physicalism, from his encyclopedia of unified science – as appears from the controversial protocol-sentence debate of the late Vienna Circle.

If we make a jump to the final phase of the Vienna Circle, we see that in turning to value issues and cultural-philosophical themes, Schlick took recourse to his ethical beginnings, which is unsurprising. In view of the fact that the backdrop of philosophy and society in Vienna at the time was the rise of Fascism and National Socialism, Schlick was prompted – like the mathematician Karl Menger – to focus on the threatening intellectual situation of his time, following his *Fragen der Ethik* (Schlick 1930).

In the midst of an antagonistic environment the democratic and liberal Schlick arrived at his 'consistent individualism' as a correlate to 'consistent empiricism' (*konsequenter Empirismus*). The omnipotent state was to be disenchanted in favour of a league of nations to promote the happiness of all people but not in the limited corset of social Darwinism:

> Accordingly, Schlick's critique of each form of nationalism as something destructive, always as a negative element with limitations, follows. Finally, he sees the National Socialist state as contrary to each form of liberal democracy (Schlick 1962, 44ff.).

It is thus no longer surprising reading in the posthumous book *Natur und Kultur* (Schlick 1952) that Schlick in the last years of his life was mainly interested in problems of cultural philosophy, along with ethics. In addition to his lectures on these issues, he also worked on a book to be called 'Nature, Culture, Art', which he was unable to finish due to his early death. The central issue of the fragment is the question as to the suffering from culture, in particular existential need, tribulations of love and the mind – of which but only the first part of the planned book was completed.

These late writings suggest Schlick was attempting a programmatic synthesis of nature and culture. It is a modern variant of a monistic world view which tries to place the realm of facts and values in a humanist and cosmopolitan context – in Schlick's own words, 'Art is a desire for nature. Culture is a bridge on both ends of which nature rests' (Schlick A 110).

Otto Neurath – encyclopedia and utopia

Otto Neurath (born 10 December 1882 in Vienna, died 22 December 1945 in Oxford), one of the last polymaths of the now lost old Austrian scientific culture, is slowly but continuously being rediscovered. This process of rediscovery is taking place on an international and interdisciplinary level although it is still a selective process and happens at varied pace and intervals. It seems not to be due merely to chance that this exercise in intellectual archaeology takes place in our era between modernism and postmodernism.

Neurath would, on principle, have rejected the title 'universal genius' which William Johnston awarded him (with the best of intentions) in his book *The Austrian Mind* (Johnston 1972). For Neurath knowledge was always a collective enterprise embedded in an all-encompassing social context. Throughout his life he never jettisoned his enlightened view of a social and cognitive totality. For him the discovery and explication of seemingly disparate connections between phenomena, both in terms of text and image and the development of lines of argument based thereon, was an essential precondition for all intellectual work. To view science as an end in itself was totally alien to him. He believed that knowledge – even as part of a hardly predictable historic process – should serve life in all the areas between the poles of everyday life and specialist research activity. On the other hand traditional academic metaphysics, obsessed with subtleties of meaning and oblivious to the consequences of material being, feels equally challenged by the radical and down-to-earth 'scientific world conception' clothed in the garb of the neo-Enlightenment, which Neurath promoted.

It is not easy to approach the complex and highly diverse life's work of Otto Neurath without, on the one hand, getting lost in details or without, on the other, resorting to inadequate generalizing descriptions such as 'positivist', 'physicalist', 'Austro-Marxist' and the like. How can one obtain an overview of a man, both of theory and practice, who covered the entire spectrum of knowledge – including architecture, economic history, mathematics and logic, philosophy, history and theory of knowledge, visual education and museology in about 300 publications (including around 30 monographs and books) – and who worked in at least four countries (Austria, Germany, the Netherlands, England) as a social scientist and teacher at commercial colleges, workers' colleges, adult education schools and universities, a man who pursued a demanding profession (as museum director), yet also figured centrally in the Vienna Circle?

It seems that the answer to this question lies in Neurath's principle of thinking and acting: to proceed from incomplete and unstable elements temporarily united in theory and its related practice and to employ, under the constraints of uncertainty and mere probability, the criteria of empiricism, without ever losing sight of the fragmentary starting point or the contemplation of a historical process. This approach in itself constitutes a criticism of the idea of a fundamental and hierarchical system and at the same time an appeal to think in ways differing from those which prevail in society and the sciences, an appeal to risk a concrete utopia and to contribute to the promotion of the common good by means of cooperative planning for freedom and happiness. After World War 1, Neurath envisaged an international 'republic of scholars', much in the tradition of the French encyclopedists. Yet despite his optimism, he realised that the growth of Fascism and National Socialism threatened to end all these young democracies, and he was not naive enough to believe that the proper proportion of rationalism and empiricism alone could reverse the 'demise of reason'. In his unpublished response to Max Horkheimer's 'Der neueste Angriff auf die Metaphysik' (1937; 'The Most Recent Attack on Metaphysics'), Neurath, anticipating the main arguments against the dialectic of Enlightenment, referred to the limitations of social criticism which result from the fragility of knowledge including sociology and its history (Neurath 1937, 2): 'What is not at all noticed at one time attracts the attention of other times and then can become the centre of important considerations [. . .] for basically nothing is secure – everything is in flux.' Moreover 'we know no other authority beyond science which sits in judgement over science and studies its foundations'; that is, there also is not available some form of transcendental reason. A central affinity to the standpoint of Wittgenstein's middle and late periods is unmistakable: philosophy is viewed as a language game, a game from which we cannot escape but about which we can certainly speak intelligently.

The basis for the remarkable rediscovery of Otto Neurath was laid some decades ago; in the English speaking world the research began (motivated not least by Willard Van Orman Quine's repeated reference to the ship metaphor as an image of knowledge) with the edition of two selected volumes of Neurath's writings (1973 and 1983), whereas in the German-speaking world, besides one volume of selected works and a monograph, with the volumes of an edition of his writings – first those on philosophy and methodology (1981), then those on visual education (1991) and now those on his life and work and on economics (1996).

In Austria an exhibition on his life and work, together with a centenary symposium on Neurath (in conjunction with that on Moritz Schlick), held in 1982, augmented the new interest by the public and scholars. Topics which attracted this rediscovery ranged from his educational work to his economic theories, his links to the architecture of the *Bauhaus* and the Viennese *Werkbundsiedlung* and to constructivist art (Haller 1982; Stadler 1982). A second main focus of interest moved towards the critical exploration of Neurath's theoretical potential and led to a reconstruction of the 'forgotten' Vienna Circle (Uebel 1991, 1992, 2007). As a result of all these activities, we can now state that the depiction by others and even self-descriptions of 'logical positivism' after 1945 – especially of Neurath's role – were derived from appalling inadequacies and prejudices. For many innovations of current history and philosophy of science were, in fact, anticipated in Neurath's oeuvre. The rediscovery of Neurath is therefore not merely a phenomenon of academic nostalgia, but itself constitutes research into the conditions and possibilities of changing a paradigm in the philosophy of science (cf. recently Cat 2010; Nemeth and Stadler 2010).

But the process of reassessment did not occur only within the narrow confines of philosophy and the theory of knowledge – the most abstract structural context of Neurath's various subjects which, on first glance, seem all too disparate in ranging from the *Encyclopedia of Unified Science* to the picture language and his museology and visual education. From the start, Neurath himself had underlined the essential unity of his patterns of thought and of action. Not only are holism in philosophy of science (Duhem-Neurath-Quine principle) and naturalism extending to the 'orchestration of the science by the encyclopedism of logical empiricism' (Neurath 1945–46) and their precise formulation pursued as systematically as his interest in visual education and the Viennese method of visual statistics and the Isotype (abbreviation for 'International System of Typographical and Pictorial Education'). Loose but deliberate analogies can be found in Neurath's original writings, as can attempts to apply his theoretical concepts to education, social reform and the politics of knowledge. The following areas must be seen from this point of view: economy in kind, war economics, planning theory, as well as his involvement in the Bavarian revolution, the Viennese housing movement and town planning projects in England, not to mention his visual education projects and the founding of the Social and

Economic Museum in Vienna and of the International Foundation for Visual Education in Holland (The Hague). Thus Neurath also emphasized several times, in a reference to Leibniz and Comenius, that the future encyclopedia did not aim merely at a standardization of the *language of science* but also at a standardization of the *visual representation* (he projected 26 volumes of text and a 10-volume 'visual thesaurus'). This combination of text with visual language had already been realized in part in his publications *Fernunterricht* (1931–3) and his successful book *Modern Man in the Making* (1939). Towards the end of his life Neurath referred to the 'mosaic of the sciences'. In the spirit of this formulation we can arrive at an understanding of his life's work by means of a kind of collage, employing the regulative idea of the unity of science and society – a term that is most appropriately used as the subtitle to the latest volume of selections from his work. (Neurath and Nemeth 1994). Neurath's phrase a 'mosaic of sciences' also contains (long before C. P. Snow's well-known critique) an implied criticism of distinct (second and third) scientific cultures.

The biography of Otto Neurath is marked by the historical caesuras of 1918–19, 1934 and 1938–40, which brought major interruptions to his work. The fact that he nevertheless carried on between destruction and reconstruction under the most trying circumstances can offer encouragement and indicate that, even in a hostile environment, the humanization of knowledge can make a modest contribution to the democratization of society (Nemeth and Stadler 1996).

First, there is Neurath's vision of a joint reconstruction of the enterprise of sciences, which, poetically formulated, reappears in different variations throughout his work.

> We are like sailors who on the open sea must reconstruct their ship but are never able to start afresh from the bottom. Where a beam is taken away a new one must at once be put there, and for this the rest of the ship is used as support. In this way, by using the old beams and driftwood the ship can be shaped entirely anew, but only by gradual reconstruction. (Neurath 1921, 199)

Then there is the ethos with which Neurath pursued this reconstruction. In the middle of the revolutionary post-war phase, Neurath delivered an analysis of the zeitgeist which, particularly in our present times of change, gives us reason to reflect:

> How often has the word 'utopian' been said with a tone of soft disdain or com-miserating recognition! However stood on the ground of what happens to be the present and only knows it, sat in judgement over he who tried to glance behind the curtain of the stage of fate, who went to pains to prepare the game of the future. Now the curtain is being raised, clumsily the actors emerge from behind the scenes. Unknown figures meet unknown figures and unknown things. Some try to feign ease, others look at the prompt boxes and listen gratefully to the utterances coming from there. Rathenau, Atlanticus, Popper-Lynkeus and others are suddenly no longer simply 'utopians', but social engineers who were ahead of their times. From all corners we now hear catchwords and demands which we find in Fourier, Cabet, Bellamy, even in Thomas Morus or Plato. The utopia today are the only attempts at total constructions of social engineering at our disposal. Whoever does not want to meet completely unarmed the innumerable stimuli and endeavours breaking over us today, should read the utopia which until now has been dealt with only marginally by economics. (Neurath 1919, 137f.)

Finally, there is the consistency and continuity of his thought. In his intellectual testament, where Neurath states:

> I always promoted monism as a means of empiricist communication and I promoted pluralism as an attitude in making hypotheses. I have contended with thinkers of all kinds who tried to declare one system as marked out before others and therefore I tried to convey the insight that we need a kind of 'decision' wherever we have to make a 'choice,' even when we are trying a scientific theory. [. . .] One may evolve more than one theory of light starting from the same basis, as one may plan more than one holiday tour from the same starting point. (Neurath 1946a, 526f.)

Rudolf Carnap – from logic of science to philosophy of science

Rudolf Carnap was born May 18, 1891 in Ronsdorf (in northwest Germany). He attended the humanist gymnasium in Barmen (today a part of the city of Wuppertal) and studied philosophy, mathematics and physics in Jena and Freiburg under Gottlob Frege, among others, from

1910 to 1914. The young Carnap participated actively in the German youth movement and served during World War I as soldier and physicist (1914–17), an experience which made him a pacifist and socialist intellectual. In 1921 he finished his dissertation on the topic of space (*Der Raum*) under the Neo-Kantian philosopher Bruno Bauch. Subsequently, Carnap continued his studies in Jena, with Hans Reichenbach among others, until 1926, before he was invited by Moritz Schlick to come to Vienna from 1925 on, where he became one of the most important members of the Vienna Circle and in logical empiricism till his emigration to the United States, where he continued his later academic career in Chicago and Los Angeles.

In 1926 he accomplished his habilitation with *The Logical Structure of the World* (published in 1928), which enabled him to become a private lecturer for theoretical philosophy at the Department of Philosophy of the University of Vienna (1926–30). In the following two years he was promoted to an associate professorship in Vienna and afterwards appointed associate professor for natural philosophy at the Faculty of Natural Science at the German University in Prague (1931–35). He was named full professor in 1936 while on leave as visiting professor at Harvard University. For scholarly and political reasons, Carnap, through the mediation of his friends Charles Morris and Willard Van Orman Quine, in 1936 emigrated permanently to the USA, where he assumed American citizenship in 1941. He was professor of philosophy at the University of Chicago (1936–52), then visiting professor at the Institute for Advanced Study in Princeton (1952–54). From 1954 to his retirement he was the successor of Hans Reichenbach at the University of California, Los Angeles, where he died on 14 September 1970 (on Carnap's life and work, see Carus 2007 and Friedman and Creath 2007).

The emergence of the discipline known today as philosophy of science can be seen as converging with the process of the increasingly scientific status of philosophy, the so-called rise of scientific philosophy (Reichenbach 1951), in the interwar years. Already in the programmatic text of the Vienna Circle – *Wissenschaftliche Weltauffassung* – *Der Wiener Kreis* (1929) – the autonomous regal discipline of philosophy had given way to an anti-metaphysical, physicalist, unified science. This idea was systematically elaborated in the thirties, most notably in Rudolf Carnap's writings. In the manifesto, reference had been primarily made to his *Logical Structure of the World* (1928) – as a constitutive system based on experience with logical analysis. A few years later the position he took in his *Logical Syntax* (1934) found acceptance. The task of *Wissenschaftslogik* is seen as lying in the study of science as a whole or in its disciplines:

> The concepts, propositions, proofs, theories appearing in the various realms of science are analyzed – less from the perspective of the historical development of science or of the sociological and psychological conditions of its functioning, but more from a logical perspective. This field of work for which no generic term has been able to gain acceptance, could be called theory of science or to be more precise logic of science. Science is understood as referring to the totality of accepted propositions. This does not just include the statements made by scholars but also those of everyday life. There is no clear boundary line drawn between these two areas. (Carnap 1934b, 5)

Here the distancing from traditional philosophy becomes highly salient, even if the role and function of a scientific *philosophy*, as linguistic analysis in Wittgenstein's sense, is not called into question. This new discipline is not so interested in propositions on the external world as the realm of the empirical disciplines – 'thing language', as in 'science itself as an orderly structure of propositions', known as object language (ibid., 6) – accordingly, in the 'sense' of the propositions and the 'meaning' of concepts from a logical point of view. The realm of these concepts is limited either to the analytic propositions of logic/mathematics or to the empirical propositions of the sciences. This culminates in the view 'that the propositions of the logic of science are propositions of the logical syntax of language. Thus these propositions lie within the boundaries drawn by Hume, for logical syntax is . . . nothing other than mathematics of language' (ibid.).

Carnap had combined the elaboration of this program of unified science in his *Logical Syntax of Language* (1934b) with its dissemination. As part of the internationalization of the Vienna Circle under way since 1929, two small books appeared almost at the same time in England: *The Unity of Science* (1934c) and *Philosophy and Logical Syntax* (1935) in the series Psyche Miniatures published by Kegan Paul. The former was an edition of the German article on physical language (Carnap 1931b), reworked by the author and translated by Max Black. The latter united three lectures that Carnap had given at the University of London in October of 1934: 'The Rejection of Metaphysics', 'Logical Syntax of

Language', 'Syntax as the Method of Philosophy'. These attempts to popularize logic of science in the Anglo-Saxon world were continued with the translation of *Logical Syntax,* which appeared in 1937 in an expanded edition from the same English publisher (Carnap 1937).

It is known that, already in his *Logical Syntax,* Carnap had been influenced by Polish and American logicians and philosophers of science (notably Tarski, Quine and Morris) to further develop the possible field of logic of science. In addition to the syntactic dimension, he cited the semantic and pragmatic dimensions as future fields of work. Accordingly, he described the logic of science in his preface to the second edition as the 'analysis and theory of the language of science':

> According to the present view, this theory comprises, in addition to logical syntax, mainly two further fields, i.e., semantics and pragmatics. Whereas syntax is purely formal, i.e., only studies the structure of linguistic expression, semantics studies the semantic relationship between expressions and objects or concepts; . . . [p]ragmatics also studies the psychological and sociological relations between persons using the language and the expressions. (Carnap 1937, vii)

With this new conceptualization of the logic of science, which already took place before the transfer of these ideas to the United States, we have also outlined the logical space for the philosophy of science as well as the terminological structure for the unity of science movement (1934c). Of course, logical empiricism before 1938 had no codified understanding of 'logic of science' in relation to philosophy. Here, however, only those paradigmatic elements have been indicated which turn out to be relevant later in the Anglo-American realm. In this context, I cannot dwell on the controversial protocol-statement debate within the Vienna Circle in which various positions on the basic issue of knowledge were unearthed. This eventually led to a heated discussion on fundamental questions in the epistemology of that time (Uebel 1992; 2007).

The movement of philosophical ideas between the old continent and the United States is meanwhile well documented (Giere and Richardson 1996; Hardcastle and Richardson 2003).

The historian of science Gerald Holton, who played a seminal role in the forties in the Unity of Science Institute and as an assistant to Philipp Frank, has given a very apt reconstruction of these cognitive parallels and this transfer of knowledge in his 'From the

Vienna Circle to Harvard Square: The Americanization of a European World Conception' (1993). This history of ideas, which also includes Quine, describes a growing internationalization best illustrated by the International Congresses and the *Encyclopedia of Unified Science* and the Unity of Science Institute founded by Frank. Holton characterizes the favourable conditions for logical empiricism in the United States from 1940 to 1969 metaphorically as an 'ecological niche' in the New World and depicted these developments as an osmotic success story.

Proceeding from the early forties as the beginning of the specific American philosophy of science, it is possible to reconstruct the intellectual conditions of the convergent development of central European and American philosophy of science (Stadler 2004, 227ff.).

In a contemporary *Dictionary of Philosophy* (Runes 1944), we find the relevant discussions of that time presented in various short entries. Here it becomes clear that the central contributions on the philosophy of science were written by Rudolf Carnap, Carl G. Hempel and Heinrich Gomperz. Carnap presents philosophy of science as

> that philosophic discipline which is the systematic study of the nature of science, especially of its methods, its concepts and presuppositions, and its place in the general scheme of intellectual disciplines. No very precise definition of the term is possible since the discipline shades imperceptibly into science, on the one hand, and into philosophy in general, on the other. A working division of its subject-matter into three fields is helpful in specifying its problems, though the three fields should not be too sharply differentiated or separated. (Carnap 1944, 284)

According to Carnap the three fields addressed here are the following:

> 1. A critical study of the method or methods of the sciences, of the nature of scientific symbols, and of the logical structure of scientific symbolic terms. . . . 2. The attempted clarification of the basic concepts, presuppositions and postulates of the sciences, and the revelation of the empirical, rational, or pragmatic grounds upon which they are presumed to rest. . . . 3. A highly composite and diverse study which attempts to ascertain the limits of the special sciences, to disclose their interrelations one with another, and to examine their implications so far as these contribute to a theory either of the universe as a whole or of some aspect of it. (Carnap 1944, 284f.)

In a preceding section, Carnap had already subsumed today's science studies under 'science of science' as 'the analysis and description of science from various points of view, including logic, methodology, sociology, and history of science' (ibid.). In this connection he referred to his entries 'Scientific Empiricism' and the 'Unity of Science' as 'a wider movement, comprising besides Logical Empiricism other groups and individuals with related views in various countries' (ibid., 286).

The unity of science was also identified with internationalization, and 'scientific empiricism' was introduced as a transformation of logical empiricism. With this self-understanding, the institutionalization and further differentiation of philosophy of science took place – a development which had been anticipated by two decades of intellectual exchange between Europe and America.

The contact with Morris gradually enabled Carnap to emigrate to the United States. After a stay in London in 1934, Carnap travelled to the United States for the first time in December 1935 – this move was also motivated by the increasingly unbearable political atmosphere in Prague. Already the year before he had met Willard Van Orman Quine (Harvard) in Vienna and Prague, which was followed by an intense dialogue and continuous contact following his emigration to Chicago in 1936.

At the University of Chicago, Carnap and Morris held a regular colloquium, known as the Chicago Circle, on methodological and interdisciplinary issues, even if the knowledge of modern logic was somewhat limited there. With this development, Carnap broke with the original conception of the logic of science (*Wissenschaftslogik*) understood as a logical syntax of language. Influenced by the work of Alfred Tarski, who immigrated from Warsaw in 1939, Carnap had undergone a 'semantic turn' in the USA by the time his *Introduction to Semantics* (1942) appeared. And the discussion of Quine's 'Two Dogmas of Empiricism' (1951) drew from the beginning on Carnap's sensitivity to the question of the analytic/synthetic or theoretical/empirical dualism.

Carnap writes about this new circle in exile:

> In Chicago Charles Morris was closest to my philosophical position. He tried to combine ideas of pragmatism and logical empiricism. Through him I gained a better understanding of the Pragmatic philosophy, especially of Mead and Dewey. For several years in Chicago we had a colloquium, founded by Morris, in which we discussed questions of methodology from scientists from various

fields of science and tried to achieve a better understanding among representatives of different disciplines and greater clarity on the essential characteristics of the scientific method. We had many stimulating lectures; but, on the whole, the productivity of the discussions was somewhat limited by the fact that most of the participants . . . were not sufficiently acquainted with logical and methodological techniques. (Carnap 1963, 34f.)

Karl Menger also participated in that circle. Referring to these meetings the editors of Karl Menger's *Reminiscences* add the following remarks:

The one tangible accomplishment of the Chicago Circle was to get some of its participants to write, and the University of Chicago Press to publish, the first monographs in the series called the International Encyclopedia of Unified Science. Apart from this the Circle suffered from an early series of blows from which, although it continued to meet in a desultory fashion until the 70's, it never fully recovered. The first of these was the departure of the noted linguist Leonard Bloomfield from the University of Chicago to become Sterling Professor at Yale. . . . The next major and practically fatal blow . . . was the war, which in the United States began in 1941, and which disrupted academic life in general. (Menger 1994, xiii–xiv).

The last two Congresses for the Unity of Science in Harvard and Chicago functioned as a forum for the transfer of knowledge and the transformation of philosophy of science into the international unity of science movement: 'Quine wrote simply: "Basically this was the Vienna Circle, with accretions, in international exile." One might say that Mach's spirit had found a resting place in the New World at long last, and that the advance of the Vienna Circle had arrived at Harvard Square' (Holton 1993, 62).

In summary, it is clear that a basis for dialogue between Vienna and Chicago in philosophy of science had been created on various levels already prior to the outbreak of World War II and the preceding cultural exodus from Austria. A path had been paved for the actual transfer of knowledge in the context of (direct and indirect) contacts, journals and congresses: the International Congresses for the Unity of Science (1935–41).

From 1938 on, publications on these activities were edited by Neurath, Carnap and Morris as part of the *International Encyclopedia of Unified Science* (IEUS), a modernist project that extended into the

1960s but was to remain uncompleted. At the same time, the journal *Erkenntnis*, edited by Carnap and Reichenbach, became international with the eighth (and last) volume as *Journal of Unified Science*, after it had come under pressure by the Nazi regime in 1933 (Spohn 1991). In 1938, the first volume of IEUS, with contributions by Neurath, Niels Bohr, John Dewey, Bertrand Russell and Carnap, marked the beginning of the uncompleted project, with 19 instead of 260 projected monographs published with the University of Chicago Press (reprint of all 19 monographs in Neurath, Carnap and Morris 1971).

Even though the editors had very different ideas about the unification of the sciences, the project was continued after the war, although the death of Neurath (1945) and the onset of the Cold War resulted in the deterioration of the whole enterprise of 'late Enlightenment'. The last path-breaking contribution by Thomas Kuhn, *The Structure of Scientific Revolutions* (1962), can be seen as reflecting a change in the philosophy of science, characterized as a pragmatic or sociological turn embedding philosophy of science in the historical context. It is really remarkable that Carnap expressed his appreciation of Kuhn's article in two letters (12 April 1960 and 28 April 1962) as part of the *Encyclopedia* – a fact, which was obscured by subsequent historiography for many reasons to be discussed in a different context (Reisch 2004).

Given the prehistory, it is no surprise that the Advisory Committee of IEUS documents a strong UK/US–Austrian bias. Accordingly, it becomes difficult to speak of an input-output or loss-gain transfer caused by the forced 'cultural exodus from Austria' from 1938 (Stadler and Weibel 1995). We are dealing more with a multilateral dynamic of science as transfer, transformation from central Europe to Great Britain and America, which can be described as a parallel process of disintegration and internationalization.

Summary

(Post-)modern philosophy of science has been strongly influenced by the direct and indirect contributions of logical empiricism (the Vienna Circle around Moritz Schlick and the Berlin Group around Hans Reichenbach) including its critics (Ludwig Wittgenstein, Karl Popper).

From the beginning of the twentieth century, we can reconstruct a long-term transfer, transformation and interaction of central European philosophy of science to the Anglo-Saxon world, from *Wissenschaftslogik* (Carnap) to philosophy of science and back to the (analytic) *Wissenschaftstheorie*. This significant development, brought on by the forced emigration of logical empiricists in the Nazi era, manifests the destruction of a creative network of philosophy of science, as well as the intense interaction of scientific philosophy and philosophy of science in central Europe (including the forgotten 'French connection' that existed with Pierre Duhem and Henri Poincaré) with the scientific community in Great Britain and North America (from the 1930s to the 1960s) – as represented by neo-pragmatism and operationalism, which centred around Percy W. Bridgman, Willard Van Orman Quine and Charles Morris, as well as linguistic and scientific philosophy (Bertrand Russell, Susan Stebbing, Frank P. Ramsey, Max Black). A critical reconstruction of today's history of philosophy of science, including exile studies and history of science, highlights this transatlantic movement and theory dynamics culminating in the long neglected re-transfer of analytic philosophy (of science) back to its roots with the 'third' Vienna Circle (around Viktor Kraft, with Arthur Pap, Paul Feyerabend and Wolfgang Stegmüller). Following the 'linguistic turn', the pragmatic and historical turns in recent philosophy of science (with Quine and Kuhn) constitute an essential part of these developments in the period from hot to cold war.

Note

This article draws significantly on my entries on the Vienna Circle and Moritz Schlick in the *Routledge Encyclopedia of Philosophy* and in *The Philosophy of Science: An Encyclopedia* (edited by Sahotra Sarkar and Jessica Pfeiffer; London: Routledge, 2006), as well as on my contributions on Otto Neurath in the volume *Otto Neurath – Encyclopedia and Utopia* (edited by Elisabeth Nemeth and Friedrich Stadler; Dordrecht, Boston and London: Kluwer, 1996) and on my 'History of the Philosophy of Science' in *Handbook of the Philosophy of Science: General Philosophy of Science – Focal Issues* (edited by Theo Kuipers; Amsterdam: Elsevier, 2007). The general background can by found in my *The Vienna Circle: Studies in the Origins, Development, and Influence of Logical Empiricism* (Vienna and New York: Springer, 2001) and in the related publications of the Institute Vienna Circle (www.univie.ac.at/ivc). Thanks go to Camilla Nielsen and Christoph Limbeck-Lilienau (both in Vienna) for their help.

References

Blumberg, Albert and Herbert Feigl (1931), 'Logical Positivism. A New Movement in European Philosophy', *Journal of Philosophy*, 28: 281–96.

Bridgman, Percy (1927), *The Logic of Modern Physics*. New York: Macmillan.

Carnap, Rudolf (1922), *Der Raum. Ein Beitrag zur Wissenschaftslehre*. Berlin: Reuther and Reichart.

— (1928), *Der Logische Aufbau der Welt*. Berlin: Weltkreis Verlag. Trans. (1967) as *The Logical Structure of the World*. Berkeley and Los Angeles: University of California Press.

— (1931), 'Überwindung der Metaphysik durch logische Analyse der Sprache', *Erkenntnis*, 2: 219–41.

— (1934a), *Die Aufgabe der Wissenschaftslogik*. Vienna: Gerold.

— (1934b), *Logische Syntax der Sprache*. Vienna: Springer. Trans. (1937) as *The Logical Syntax of Language*. London: Routledge.

— (1934c), *The Unity of Science*. London: Kegan Paul.

— (1935), *Philosophy and Logical Syntax*. London: Kegan Paul.

— (1936), 'Von der Erkenntnistheorie zur Wissenschaftslogik', in *Actes de Congrès International de Philosophie Scientifique* I. Paris: Hermann, 36–41.

— (1942), *Introduction to Semantics*. Cambridge, MA: Harvard University Press.

— (1944), Entries in *The Dictionary of Philosophy*, Dagobert Runes (ed.). London: Routledge.

— (1963), 'Intellectual Autobiography', in *The Philosophy of Rudolf Carnap*, Paul A. Schilpp (ed.). La Salle, IL: Open Court, 3–84.

Carnap, Rudolf, Hans Hahn and Otto Neurath (1929), *Wissenschaftliche Weltauffassung. Der Wiener Kreis*, Verein Ernst Mach (ed.). Vienna: Artur Wolf Verlag. Abridged and trans. as 'The Scientific Conception of the World: The Vienna Circle', in Neurath (1973), 299–318.

Cartwright, Nancy, Jordi Cat, Lola Fleck and Thomas Uebel (eds) (1996), *Otto Neurath: Philosophy Between Science and Politics*. Cambridge: Cambridge University Press.

Cat, Jordi (2010), 'Neurath, Otto', in *Stanford Encyclopedia of Philosophy*, http://plato.stanford.edu.

Einstein, Albert (1921), *Geometrie und Erfahrung*. Berlin: 1921. Trans. (1922) as *Sidelights on Relativity*. London, New York: Kessinger Publishing.

Fischer, Kurt and Friedrich Stadler (eds) (1997), *'Wahrnehmung und Gegenstandswelt'. Zum Lebenswerk von Egon Brunswik (1903–1955)*. Vienna and New York: Springer.

Frank, Philipp (1949), *Modern Science and Its Philosophy*. Cambridge, MA: Harvard University Press. See esp. the Introduction – Historical Background).

Friedman, Michael and Richard Creath (eds) (2007), *The Cambridge Companion to Carnap*. Cambridge: Cambridge University Press.

Giere, Ronald N. and Alan W. Richardson (eds) (1996), *Origins of Logical Empiricism*. Minneapolis and London: University of Minnesota Press.

Haller, Rudolf (ed.) (1982), *Schlick und Neurath – Ein Symposion* (Grazer Philosophische Studien 16/17). Amsterdam: Rodopi.

Hardcastle, Gary and Alan W. Richardson (eds) (2003), *Logical Empiricism in North America*. Minneapolis: University of Minnesota Press.

Heidelberger, Michael and Friedrich Stadler (eds) (2003), *Wissenschaftsphilosophie und Politik*, trans. as *Philosophy of Science and Politics*. Vienna and New York: Springer.

Hempel, Carl, G. (1993), 'Empiricism in the Vienna Circle and in the Berlin Society for Scientific Philosophy. Recollections and Reflections', in Stadler (1993), 1–10.

Holton, Gerald (1993), "From the Vienna Circle to Harvard Square: The Americanization of a European World Conception", in Stadler (ed.), 47–74.

Horkheimer, Max (1937), 'Der neueste Angriff auf die Metaphysik,' *Zeitschrift für Sozialforschung*, 1937, VI/1.

Johnston, William (1972), *The Austrian Mind: An Intellectual and Social History 1848–1938*. Berkeley: University of California Press, ch. 12.

Kaufmann, Felix (1936), *Methodenlehre der Sozialwissenschaften*. Vienna: Springer. Reprint (1999), Vienna and New York: Springer.

Kuipers, Theo (ed.) (2007), Handbook of the Philosophy of Science: General Philosophy of Science – Focal Issues. Amsterdam: Elsevier.

Manninen, Juha and Friedrich Stadler (eds) (2010), *The Vienna Circle in the Nordic Countries: Networks and Transformations of Logical Empiricism*. Vienna and New York: Springer.

Marcuse, Herbert (1964), The One-Dimensional Man: Studies in the Ideology of Advanced Industrial Society. Boston: Beacon Press.

McGuinness, Brian (ed.) (1985), *Zurück zu Schlick. Eine Neubewertung von Werk und Wirkung*. Vienna: Hölder-Pichler-Tempsky.

Menger, Karl (1979), *Selected Papers in Logic and Foundations, Didactics, Economics*. Dordrecht, Boston and London: Reidel.

— (1994), *Reminiscences of the Vienna Circle and the Mathematical Colloquium*, Louise Golland, Brian McGuinness and Abe Slar (eds). Dordrecht, Boston and London: Kluwer.

Naess, Arne (2003), 'Pluralism of Tenable Worldviews', in Stadler (2003a).

Nemeth, Elisabeth and Friedrich Stadler (eds) (1996), *Encyclopedia and Utopia: The Life and Work of Otto Neurath (1882–1945)*. Dordrecht, Boston and London: Kluwer.

— (forthcoming 2011), 'Workshop on Otto Neurath's Visual Language and Pictorial Statistics'. Proceedings of the International Wittgenstein Symposium 2010.

Neurath, Otto (1919), 'Utopien', in *Wirtschaft und Lebensordnung*. Dresden. Reprinted (1981) as *Gesammelte philosophische und methodologische Schriften*, 137f.

— (1921), *Anti-Spengler*. Munich: Callwey. Trans. in Neurath (1973), 158–213.

— (1931), Empirische Soziologie. Der wissenschaftliche Gehalt der Geschichte und Nationalökonomie. Vienna: Springer. Trans. in Neurath (1973), 319–421.

— (1937), 'Einheitswissenschaft und Logischer Empirismus. Eine Erwiderung.' Unpublished MS. Vienna Circle Foundation, Rijksarchief Noord Haarlem (NL), 2.

— (1939) *Modern Man in the Making*. New York: Alfred Knopf.

— (1946a), 'For the Discussion: Just Annotations, Not a Reply', *Philosophy and Phenomenological Research*, VI (1946): 526f.

— (1946b), 'The Orchestration of the Sciences by the Encyclopedia of Logical Empiricism', *Philosophy and Phenomenological Research*, VI/4: 496–508.

— (1973), *Empiricism and Sociology*, Marie Neurath and Robert S. Cohen (eds). Dordrecht and Boston: Reidel.

— (1981), *Gesammelte philosophische und methodologische Schriften* (GPMS), 2 vols, Rudolf Haller and Robin Kinross (eds). Vienna: Hölder-Pichler-Tempsky.

— (1983), *Philosophical Papers 1913–1946*, Robert S. Cohen and Marie Neurath (eds). Dordrecht, Boston and Lancaster: Reidel.

— (1991), *Gesammelte bildpädagogische Schriften* (GBS), Rudolf Haller and Robin Kinross (eds). Vienna: Hölder-Pichler-Tempsky.

— (2004), *Economic Writings. Selections 1904–1945*, Thomas E. Uebel and Robert S. Cohen (eds). Dordrecht, Boston and London: Kluwer.

Neurath, Paul and Elisabeth Nemeth (eds) (1994), *Otto Neurath, oder die Einheit von Wissenschaft und Gesellschaft*. Vienna, Cologne and Weimar: Böhlau.

Neurath, Carnap and Morris, Charles (1971), Foundations of the Unity of Science. Toward an International Encyclopedia of Unified Science. Ed. by Otto Neurath, Rudolf Carnap, Charles Morris. Chicago and London: The University of Chicago Press. 2 vols. 1970/71.

Quine, Willard Van Orman (1951), 'Two Dogmas of Empiricism', *Philosophical Review*, 60: 20–43. Reprinted (1953) in *From a Logical Point of View: Nine Logico-Philosophical Essays*. Cambridge, MA: Harvard University Press.

Reichenbach, Hans (1951), *The Rise of Scientific Philosophy*. Berkeley: University of California Press. Publ. in German as *Der Aufstieg der wissenschaftlichen Philosophie*. Berlin-Grunewald: Herbig.

Reisch, George (2004), How the Cold War Transformed Philosophy of Science: To the Icy Slopes of Logic. Cambridge: Cambridge University Press.

Richardson, Alan and Thomas Uebel (eds) (2007), *The Cambridge Companion to Logical Empiricism*. Cambridge: Cambridge University Press.

Schlick, Moritz (1908), Lebensweisheit. Versuch einer Glückseligkeitslehre. Munich.

— (1917), 'Raum und Zeit in der gegenwärtigen Physik. Zur Einführung in das Verständnis der allgemeinen Relativitätstheorie', in *Die Naturwissenschaften*, 5,

161–67, 177–86. Publ. in English (1920) as *Space and Time in Contemporary Physics*, Henry L. Brose (trans.). Oxford and New York.

— (1918–25), *Allgemeine Erkenntnislehre*. Berlin: Springer. Publ. in English (1974) as *General Theory of Knowledge*, Albert E. Blumberg (trans.), with an introduction by A. E. Blumberg and Herbert Feigl. Vienna and New York: Springer.

— (1930), *Fragen der Ethik*. Vienna: Springer. Publ. in English (1962) as *Problems of Ethics*, David Rynin (trans.). New York: Prentice-Hall.

— (1932–3), 'Positivismus und Realismus', *Erkenntnis*, 3: 1–31. Trans. as 'Positivism and Realism', *Synthese* 7: 478–505.

— (1934), 'Über das Fundament der Erkenntnis', *Erkenntnis*, 4: 79–99.

— (1950), 'Moritz Schlick', in *Philosophen-Lexikon*, Werner Ziegenfuss and Gertrud Jung (eds). Berlin: de Gruyter.

— (1952), *Natur und Kultur*, aus dem Nachlaß, Josef Rauscher (ed.). Vienna: Gerold.

— (1979), *Philosophical Papers*, 2 vols, Henk L. Mulder and Barbara van de Velde-Schlick (eds), Peter Heath (trans.). Dordrecht, Boston and London: Reidel.

— (2005ff.), Moritz Schlick Edition Project. Critical Edition of the Complete Works and Intellectual Biography. Moritz Schlick Studies. Friedrich Stadler (Vienna) and Hans-Jürgen Wendel (Rostock) (general eds). Vienna and New York: Springer. www.univie.ac.at/ivc/Schlick-Projekt/, www.moritz-schlick.de.

Smith, Laurence D. (1986), *Behaviorism and Logical Positivism: A Reassessment of the Alliance*. Palo Alto, CA: Stanford University Press.

Spohn, Wolfgang (1991), Erkenntnis Oriented: A Centennial Volume for Rudolf Carnap and Hans Reichenbach. Dordrecht et al.: Kluwer.

Stadler, Friedrich (ed.) (1982), Arbeiterbildung in der Zwischenkriegszeit. Otto Neurath – Gerd Arntz. Vienna and Munich: Löcker.

— (ed.) (1987–88), *Vertriebene Vernunft. Emigration und Exil österreichischer Wissenschaft*, 2 vols. Vienna and Munich: Jugend und Volk; 2nd edn (2004), Münster: LIT Verlag.

— (ed.) (1988), Kontinuität und Bruch 1938–1955. Beitrage zur österreichischen Kultur- und Wissenschaftsgeschichte. Vienna and Munich: Jugend und Volk; 2nd edn (2004), Münster: LIT Verlag.

— (ed.). (1993), *Scientific Philosophy: Origins and Developments*. Dordrecht, Boston and London: Kluwer.

— (2001), *The Vienna Circle. Studies in the Origins, Development, and Influence of Logical Empiricism*. With the first publication of the protocols (1930–1) of the Vienna Circle and an interview with Sir Karl Popper (1991). Vienna and New York: Springer.

— (ed.) (2003a), The Vienna Circle and Logical Empiricism. Reevaluation and Future Perspectives. Dordrecht, Boston and London: Kluwer.

— (2003b), 'The 'Wiener Kreis' in Great Britain: Emigration and Interaction in the Philosophy of Science,' in Timms and Hughes (2003), 155–80.

— (2003c), 'Transfer and Transformation of Logical Empiricism: Quantitative and Qualitative Aspects,' in Hardcastle and Richardson, 216–33.

— (2007), 'History of the Philosophy of Science. From *Wissenschaftslogik* (Logic of Science) to Philosophy of Science: Europe and America, 1930–1960', in Kuipers (2007), 577–658.

— (ed.) (2010a), *The Present Situation in the Philosophy of Science*. Dordrecht, Heidelberg, London and New York: Springer.

— (ed.) (2010b), Vertreibung, Transformation und Rückkehr der Wissenschaftstheorie. Am Beispiel von Rudolf Carnap und Wolfgang Stegmüller. Vienna and Berlin: LIT Verlag.

Stadler, Friedrich and Peter Weibel (eds) (1995), *The Cultural Exodus from Austria*. Vienna and New York: Springer.

Timms, Edward and Jon Hughes (eds) (2003), Intellectual Migration and Cultural Transformation: Refugees from National Socialism in the English-Speaking World. Vienna and New York: Springer.

Uebel, Thomas (ed.) (1991), Rediscovering the Forgotten Vienna Circle: Austrian Studies on Otto Neurath and the Vienna Circle. Dordrecht, Boston and London: Kluwer.

— (1992), Overcoming Logical Positivism from Within: The Emergence of Neurath's Naturalism in the Vienna Circle's Protocol Sentence Debate. Amsterdam and Atlanta: Rodopi.

— (2000), Vernunftkritik und Wissenschaft. Otto Neurath und der Erste Wiener Kreis. Vienna and New York: Springer.

— (2007), Empiricism at the Crossroads: The Vienna Circle's Protocol-Sentence Debate. Chicago: Open Court, 2007.

Von Mises, Richard (1951), *Positivism: A Study in Human Understanding*. New York: Dover.

CHAPTER 4

CARL G. HEMPEL: LOGICAL EMPIRICIST
Martin Curd

Best known for his work on confirmation, explanation and the interpretation of scientific theories, Carl Hempel (1905–1997) was one of the twentieth century's leading philosophers of science. He was also an acute critic of logical positivism, attacking its objectionably narrow empiricist doctrines – verificationism, operationism and instrumentalism – while remaining true to the underlying spirit of analytic empiricism (which was Hempel's preferred term for his more liberal brand of logical empiricism). Universally respected for his intellect, Hempel (known to his friends as Peter) was also admired for his openmindedness and generosity of spirit. In the tributes commemorating his life, two themes stand out: the affection and respect of his colleagues, former students and friends (many people in the profession were all three); and Hempel's candour in recognizing problems for positions he defended and his willingness to change his mind. Thomas Kuhn aptly described him as 'a philosopher who cares more about arriving at truth than about winning arguments'.[1]

Life and work

Born in 1905 in Oranienburg, Germany, Hempel grew up in the suburbs of Berlin. He attended the universities of Göttingen and Heidelberg, where he studied mathematics, physics, and philosophy. From David Hilbert and others at Göttingen, Hempel developed a special interest in symbolic logic and the foundations of mathematics. In 1924 he moved to the University of Berlin, where he became part of the Berlin Circle, the Society for Empirical Philosophy, led by Hans Reichenbach.[2] Among Hempel's teachers at Berlin were Max Planck and John von Neumann, but Hempel was most influenced by Reichenbach and his courses on logic, the philosophy of space and time and the theory of probability.

Hempel was impressed by Rudolf Carnap's early writings, notably *The Logical Construction of the World* and *Pseudoproblems in Philosophy*, both published in 1928. These works seemed to Hempel, at that time, to hold the solution to all genuine philosophical problems. Hempel met Carnap in Prague in 1929 and, with Reichenbach's encouragement, Hempel then went to Vienna where he took seminars with Carnap, Friedrich Waismann and Moritz Schlick and attended meetings of the Vienna Circle, where he also interacted with figures such as Otto Neurath, Hans Hahn, Herbert Feigl and Karl Popper.

Upon his return to Berlin, Hempel began writing his dissertation (directed by Reichenbach) on the interpretation of probability statements. With the collapse of the Weimar republic, Hitler became chancellor of Germany; anti-Semitism and thuggish nationalism were now official state policy.[3] Because of the Nazis' racial laws, Reichenbach left Berlin for the University of Istanbul in 1933, later taking up a position at UCLA in 1938. Paul Oppenheim (a fellow member of the Berlin circle) left Germany for Brussels. After obtaining his doctorate from the University of Berlin in 1934, Hempel gratefully accepted an offer from Oppenheim to work with him in Belgium.[4] Their collaboration resulted in a co-authored monograph analyzing the logic of qualitative concepts, especially in constitution theory, which attempts to relate human personality to physical body type.

In 1937 Carnap invited Hempel and Olaf Helmer to the University of Chicago as his research associates for a year. Then, after a brief return to Brussels, Hempel and his wife, Eva, were again able to reach the safety of the United States. Hempel held teaching positions at City College (1939–1940) and then at Queens College (1940–1948) in New York. During this period Hempel published a series of path-breaking papers (several in collaboration with Oppenheim, who was now working as an independent scholar in Princeton, New Jersey), mainly in confirmation theory and explanation. These papers established Hempel's reputation and set the course of research in philosophy of science for several decades.

From 1948 to 1955, Hempel taught at Yale University; he left to become Stuart Professor of Philosophy at Princeton University (1955–73). After having reached the mandatory retirement age at Princeton, Hempel continued there as a lecturer for two years. In 1975 Hempel joined with Wolfgang Stegmüller and Wilhelm K. Essler in reviving *Erkenntnis*, a journal devoted to analytic philosophy and philosophy of science.[5] Following lecture tours in Australia and China, Hempel joined the Department of Philosophy at the University of Pittsburgh, first as a visitor (in 1976) and then as a university professor (1977), before retiring for a second time in 1985.

A noteworthy feature of Hempel's years at Princeton was his interaction with Thomas S. Kuhn. Hempel first met Kuhn while Hempel was a fellow at the Stanford Center for Advanced Study in the Behavioral Sciences (1963–4). At that time Kuhn was teaching history of science at Berkeley and had just published *The Structure of Scientific Revolutions* (1962).[6] While Hempel at first resisted Kuhn's ideas, he later acknowledged Kuhn as a significant influence on the evolution of his thinking about science. In 1964, Kuhn left Berkeley to become Hempel's colleague at Princeton as M. Taylor Pyne Professor of Philosophy and History of Science. Though in different departments (Kuhn was in history, Hempel in philosophy), they collaborated in Princeton's Program in the History and the Philosophy of Science. This led Hempel to publish several papers comparing his analytic empiricism with Kuhn's pragmatist and historicist approach.

Hempel has been called a philosopher's philosopher. His work presupposes a familiarity with logic and is addressed primarily to fellow professionals rather than the general public. Hempel's unit of philosophical production was the paper rather than the book. His most important papers are conveniently collected in three works: *Aspects of Scientific Explanation and Other Essays in the Philosophy of Science* (1965); *Selected Philosophical Essays: Carl G. Hempel*, Richard Jeffrey, ed. (2000) and *The Philosophy of Carl G. Hempel: Studies in Science, Explanation, and Rationality*, James H. Fetzer, ed. (2001). Hempel's books include *Fundamentals of Concept Formation in Empirical Science* (1952) and his lucid introductory textbook *Philosophy of Natural Science* (1966).

Verificationism and operationism

Hempel went to Vienna in 1929 to imbibe logical positivism from its source: the philosophers, mathematicians, logicians, physicists and social scientists belonging to the Vienna Circle. Though the Viennese positivists often disagreed on important philosophical matters, they presented a unified front in their *Manifesto* (1929) and popular lectures. Among the common themes of their official 'scientific conception of the world' were a rejection of the synthetic a priori, a commitment to the unity of science, the appeal to verifiability as the criterion of cognitive significance, logicism (the doctrine that mathematics is reducible to logic), an insistence of the distinction between logic and psychology and the replacement of traditional disputes in philosophy (especially metaphysics) with analyses in terms of language. Many of these positions and doctrines were later modified or abandoned as a result of searching criticisms by the positivists themselves (notably Hempel, Carnap and Quine) as logical positivism evolved into logical empiricism.

The logical positivists identified verifiability as a necessary and sufficient condition for 'empirical meaningfulness' or 'cognitive significance.' Only when a sentence is cognitively significant is it capable of being true or false. Unlike theology, ethics and poetry (which may have their own distinctive kind of emotive meaning but are unverifiable), we can verify or falsify the well-formed sentences of the empirical sciences because and only because of their logical connection with experience. For the logical positivists, experience is expressed linguistically in the form of observation statements.

Like most philosophers after the 1940s, Hempel (1951, 1952) rejected the full-blooded form of the verifiability principle (as well as many weaker versions proposed to replace it). In its original formulation the verifiability principle judges a synthetic sentence to be significant if and only if it can be completely verified by deducing it from some finite, consistent set of observation sentences. As Popper and others quickly pointed out, this complete verifiability version of the principle implies that all universal generalizations (and hence all scientific laws) are non-significant. More generally, it is difficult to see how important scientific theories such as the wave theory of light and the kinetic theory of gases could be deduced from any finite set of observation sentences. This problem soon led most (but not all) of the logical positivists to abandon the conclusive verifiability principle in favour of a weaker, confirmability version.[7]

The core idea behind confirmability is that, instead of demanding that observation sentences entail a scientific law or theory, we merely require that the law or theory entails an observation sentence, and so makes at least one testable prediction that can in principle be checked. If the prediction turns out to be correct, the theory is thereby confirmed. This is the basic picture given by the hypothetico-deductive model of science.

Usually, a scientific theory, S, has to be conjoined with additional statements, E (e.g., initial conditions, boundary conditions, auxiliary hypotheses), before any observation sentence can be deduced. In that case, we need to ensure that S is essential to the derivation by stipulating that the additional statements, E, do not entail the observation statement all by themselves. Symbolically we have the following: S is cognitively significant if and only if $(S \& E)$ is consistent, $(S \& E)$ entails O_1, and $\sim(E$ entails $O_1)$.

It was soon realized that this explication of confirmability is far too weak. Even a sentence such as 'The Absolute is wonderful' would count as significant since it entails an observation sentence when conjoined with 'If the Absolute is wonderful, then this lemon is yellow.' In response to this problem, proposals were made limiting the additional statements, E, to those that are themselves confirmable, but all these efforts failed. In every case, the new proposal turned out to be too liberal, letting in as significant just about anything. Not everyone has despaired of finding further patches to repair the principle of cognitive significance but most philosophers agree with Hempel that the prospects for success are bleak.

A distinctive feature of Hempel's criticism of principles of cognitive significance is the condition of adequacy he adopts, stating that if a sentence P is nonsignificant, then so too are all truth-functional compound sentences in which P appears as a non-vacuous component.[8] Thus, if P is nonsignificant, then so too is its negation ($\sim P$) and, more generally, any compound sentences formed from P by disjunction or conjunction. Thus, according to Hempel, if P is nonsignificant, then so too is $(P \lor Q)$ and $(P \& Q)$, regardless of the status of Q. It is easy to show that complete verifiability and various confirmability principles all violate Hempel's condition. For example, while a generalization such as $(x)(Rx \supset Bx)$ is nonsignificant according to the complete verifiability principle, its negation qualifies as significant because $(\exists x)(Rx \, \& \, {\sim}Bx)$ is deducible from observation sentences. Similarly, confirmability principles judge S to be significant if observational consequences can be deduced from S in the 'right' sort of way; but those same consequences are deducible from $(S \, \& \, N)$ even though N might be 'The Absolute is wonderful,' which is nonsignificant.[9]

Given the failure of attempts to define cognitive (or empirical) significance in terms of the verifiability or confirmability of sentences, a tempting alternative is to look at a sentence's non-logical terms and see whether each of these is empirically acceptable. This strategy was elaborated by the American physicist Percy W. Bridgman in *The Logic of Modern Physics* (1928). One of its advantages (noted by Hempel) is that truth-functional complex sentences will automatically be judged nonsignificant if they contain any non-significant component. Bridgman described his approach (which he called operationism) as follows: 'In general, we mean by any concept nothing more than a set of operations; *the concept is synonymous with the corresponding set of operations.*'

Problems with Bridgman's proposal are exposed in Hempel (1954). Much depends on what one understands by a 'set of operations' and how narrowly it is specified. If it means a detailed description of a measuring process, then Bridgman's proposal would lead to a fragmentation of scientific concepts. For example, each different way of measuring temperature would be regarded as defining a different concept of temperature. This would impede the unifying goal of science and make it hard for scientists to view themselves as using new instruments and techniques to extend and refine their ability to measure a single, core concept.

Equally worrying is the dogmatism that would creep in if one were to collapse the distinction between a concept and a particular means for measuring its value. If a measuring process (e.g., a particular kind of test for IQ) is taken to specify an empirical concept *by definition*, then important questions about a particular test can no longer be raised: Does it *accurately* measure IQ? Does IQ (as measured by the test) really exist? Does a different test give a more reliable measure of IQ? The hard-core operationist will assert that 'IQ_T' is what is measured by test T, and refuse to entertain these and other questions. This seems to be an unwarranted dismissal of questions which, on the face of it, are significant and meaningful.

Another problem concerns the role of mathematical calculations in scientific measurements. Only in the simplest cases is the result of measurement obtained by recording the position of a pointer on a dial (let us say): instruments have to be calibrated, corrections made, compensations introduced for interfering factors and statistical error analysis performed on sets of data points. Bridgman allowed for the inclusion of 'pen and paper' and other 'mental' operations in the total set of techniques that would define a concept, but it is hard to limit the scope of such operations to those that are empirically respectable.

Rudolf Carnap introduced reduction sentences in his 'Testability and Meaning' (1936). He hoped to avoid the problems with Bridgman's operational definitions while still satisfying the empiricist demand that theoretical language be made meaningful by tying it to observational tests. As developed by Carnap, reduction sentences are said to provide *partial* definitions of theoretical concepts in terms of observables.

Suppose that we try to define a theoretical term T, using a test involving a stimulus (or test condition) S and a characteristic response R that objects satisfying T manifest when placed in circumstances S. Let T stand for the dispositional property of being fragile.

The test involves hitting the object of interest sharply (S) and seeing whether it breaks (R). It is assumed that S and R, unlike T, are expressed in an observation language: the empirical significance of S and R is already given; the task is to define T in terms of S and R. Suppose, in the spirit of Bridgman's operationism, we try to define T as follows:

(A): $Tx \equiv (Sx \supset Rx)$.

In English, this reads, 'Object x is fragile if and only if it breaks when it is struck sharply.' As Carnap pointed out, (A) fails to capture the meaning of the dispositional term T. The problem lies in the material conditional linking S and R. By definition, a material conditional is true whenever its antecedent is false or its consequent true. So, if object x (say a piece of rubber at room temperature) is never struck, it follows from (A) that it is fragile.[10] The problem is that the S-R test for fragility is reliable only when the stimulus condition, S, actually obtains. This suggests replacing (A) with (B).

(B): $Sx \supset (Tx \equiv Rx)$

Unlike (A), (B) does not give a single necessary and sufficient condition for applying the theoretical term T. According to (B), an object is fragile if it is struck sharply and breaks: $[(Sx \ \& \ Rx) \supset Tx]$. If the object is never struck sharply, (B) remains true but it says nothing about whether or not the object is fragile. A necessary condition for T, according to (B), is given by $[Tx \supset (Sx \supset Rx)]$. In other words, an object is fragile only if it breaks when it is sharply struck; or, equivalently, an object is not fragile if it is struck sharply and fails to break. So (B) gives us a condition under which T applies, and it gives us a different condition under which $\sim T$ applies. Thus, (B) does not give a full and complete definition of the dispositional concept T in terms of the observables S and R. For this reason, Carnap referred to open sentences such as (B) as partial definitions of theoretical terms. Reduction sentences come in many different logical forms, but none of them give complete definitions of theoretical terms. Thus, reduction sentences cannot be used to eliminate theoretical terms by replacing them with equivalent expressions solely using observables.

Hempel made two important points about reduction sentences. First, Hempel emphasized that sets of reduction sentences involving the same theoretical terms can have empirical content. This counts against them as playing the role of definitions in the way that logical empiricists regarded definitions, since anything with empirical content is not analytic. For example, the pair of reduction sentences

$[Sx \supset (Tx \equiv Rx)]$ and $[Px \supset (Tx \equiv Qx)]$

entail the empirical sentence $\sim(\exists x)(Sx \ \& \ Rx \ \& \ Px \ \& \ \sim Qx)$.

More generally, whether or not a reduction sentence is analytic or synthetic depends on what other sentences the theory includes. This dependence of analyticity on theoretical context echoes Quine's famous attack on the analytic/synthetic distinction in his 'Two Dogmas of Empiricism' (1951).

Hempel's other point is that requiring all theoretical terms to be introduced by chains of reduction sentences is too restrictive. The important point is that the derivation of testable predictions (expressed solely in observational vocabulary) from the entire theory is what secures 'empirical significance' for theoretical terms, *whether or not* this involves reduction sentences. The Ψ-function in quantum mechanics is a good example. Carnap readily conceded this point in his later writings.

Confirmation

During the 1940s Hempel worked on two aspects of confirmation theory: quantitative and qualitative. The goal of the quantitative theory was to define a numerical function $c(h, e)$ for any pair of sentences h and e, such that it gives the degree of confirmation that e confers on h.[11] Perhaps understandably, neither Carnap nor Hempel succeeded in producing a theory that could be applied to the confirmation of real-life scientific theories and hypotheses. Later in his career Hempel became increasingly sceptical that any such theory, based solely on formal logic, syntax and semantics, could do justice to the complexity of reasoning about evidence and theory-choice in science.

Qualitative theories of confirmation are classificatory. In their simplest form, they tell us when a sentence (such as a scientific hypothesis) is confirmed (or not) by an observation statement without specifying how strong the support is. In one of his best known papers, 'Studies in the Logic of Confirmation' (1945), Hempel defended the view that generalizations and laws are confirmed by their positive instances. This led directly to the so-called raven paradox, which became the focus of intense discussion for several decades.

Hempel asks us to consider the hypothesis, H_1, that all ravens are black: $(x)(Rx \supset Bx)$. He thinks it obvious that whatever confirms a hypothesis must also confirm anything that is logically equivalent to that hypothesis. The equivalent forms of H_1 include H_2: all nonblack things are nonravens, and H_3: anything that is either a raven or not a raven is either a nonraven or black.

H_1: $(x)(Rx \supset Bx)$
H_2: $(x)(\sim Bx \supset \sim Rx)$
H_3: $(x)[(Rx \lor \sim Rx) \supset (\sim Rx \lor Bx)]$

What confirms H_2? Presumably the observation of a nonblack non-raven, expressed in the observation sentence E_2: $(\sim Ba \ \& \ \sim Ra)$. But H_2 is equivalent to H_1 and so E_2 also confirms that all ravens are black. It also follows (from the equivalence of H_3 and H_1) that the observation of nonraven (whether black or not) and the observation of a black thing (whether a raven or not) also confirms that all ravens are black. In this way, observations of white shoes, lemons, and soot each confirm H_1. Though these conclusions may seem paradoxical, Hempel insists that they are correct. The key is to disabuse ourselves of the notion that H_1 is really about ravens. When expressed as a universal generalization using first-order predicate logic, H_1 literally says, 'If anything is a raven, then that thing is black.' The range of this expression could be things or even spatiotemporal regions. In the latter case, H_1 asserts that for any place in the universe, if the property of being a raven is instantiated in that place, then so too is the property of being black. From that perspective it is perhaps not so counterintuitive that H_1 is confirmed, to whatever small degree, whenever we observe a region in which H_1 is not violated. In a long footnote, Hempel acknowledges the contribution of the Polish logician Janina Hosiasson-Lindenbaum in framing this response. Hosiasson-Lindenbaum had suggested that it is because we know that ravens are vastly less abundant than nonblack things, that we regard H_1 as confirmed much more strongly by the observation of black ravens than by the observation of nonblack nonravens. In the case of nonblack nonravens, we may then mistake a very low degree of confirmation for no confirmation at all.

Later in his article Hempel gives a formal definition of what he calls the satisfaction criterion of confirmation. Simply put, a hypothesis is confirmed by an observation report if the individuals mentioned in the report satisfy the hypothesis. Satisfaction includes but is not restricted to positive instances. More technically, a hypothesis is satisfied (and hence confirmed by) an observation report if the report entails the development of the hypothesis, that is, a description of the world that would be true if the hypothesis were true and if the world contained only the individuals mentioned, essentially, in the report. Suppose, for example, that the report says that Beth is not a raven: $\sim Rb$. That report mentions Beth. The development of H_1 for a world that contains only Beth is $(\sim Rb \lor Bb)$ and that proposition is entailed by $\sim Rb$. Therefore, $\sim Rb$ confirms H_1. Similarly, the report $(Ra \& Ba)$ confirms H_1 because it entails $(\sim Ra \lor Ba)$.

It is a general feature of Hempel's approach that the extra information in a report – that is, any information that goes beyond the development of a hypothesis for the individuals mentioned in the report – makes no difference as to whether or not the hypothesis is confirmed. That assumption has been widely challenged, especially by philosophers who take a Bayesian approach to confirmation theory. Here is a simple example (adapted from Rosenkrantz 1982). Three men (Adam, Bob and Cal) go to a concert and check their hats. The hats are given back to them after the performance but they are distributed at random. Consider the hypothesis, *J*, that no man receives his own hat. If we observe that Adam receives Bob's hat, and Bob receives Adam's, the report entails that, for those two individuals, neither receives his own hat, which is the development of *J*. But the extra information in the report also entails that *J* must be false (since Cal must receive his own hat).[12]

Several aspects of Hempel's account stand out. It is restricted to theories that share the same vocabulary (observational predicates) with the reports that confirm (or disconfirm) them. Thus it might apply to low-level scientific laws and generalizations (such as Hooke's law and Boyle's law) but not to theories that postulate unobservable entities such as atoms, genes and viruses. It also assumes that scientific laws are correctly analyzed as universal generalizations so that their content can be expressed in many different but fully equivalent ways. In offering his theory Hempel explicitly rejects the hypothetico-deductive account of confirmation. In his view it is not sufficient for confirmation that one can derive a correct prediction from a theory in conjunction with statements of initial conditions, auxiliary hypotheses and the like. Thus Hempel denies that (*Ra* & *Ba*) confirms H_1 because (H_1 & *Ra*) correctly predicts *Ba*; in order to confirm H_1 a report has to be an instance of H_1 in the sense spelled out by the satisfaction criterion.

Lastly, Hempel's account is purely syntactical. It presupposes that we can judge whether or not an observation report confirms a hypothesis solely on the basis of the logical form of the relevant propositions. That last assumption was challenged by Nelson Goodman's so-called grue paradox. Hempel took Goodman's challenge very seriously. When his 'Studies in the Logic of Confirmation' paper was reprinted in *Aspects*, Hempel added a postscript in which he credited Goodman with having shown that confirmation, whether qualitative or quantitative, cannot be defined solely syntactically but must depend on pragmatic factors. Here is Hempel's version of Goodman's argument. Define the predicate 'blite' as follows: an object is blite if it has been examined before a certain time, *t*, and is black, or has not been examined before *t* and is white. Suppose that we have examined a number of ravens before *t* and found them all to be black. That entails that the observed ravens are blite. But surely observation reports of black ravens examined before *t* do not confirm the hypothesis that all ravens are blite even though, on Hempel's theory, they are positive instances of that hypothesis. Goodman proposed that whether a hypothesis is confirmed by its positive instances (whether it is 'projectable') depends on the entrenchment of its constituent predicates. That in turn is a function of how often the predicates (or predicates that have the same extension) occur in hypotheses that have been successfully projected in the past. Thus confirmation depends, in part, on the track record of our past inductive practice.

Explanation

Hempel's first paper on explanation (1942) was motivated by the desire (typical of the unity of science movement) to defend the view that explanation is fundamentally the same in both the natural and the social sciences. Unlike some of his contemporaries, Hempel rejected the notion that there is a special kind of empathy or subjective understanding (*Verstehen*) that is essential to explaining human actions. As in physics, so in history, genuine explanations are objective and essentially involve laws. In the case of historical explanations, those laws are empirical generalizations about individual and social psychology. In this way Hempel began his long defence of the 'covering law model' of explanation. Six years later, Hempel and Oppenheim (1948) laid out the basic elements of the covering law model, focusing on deductive-nomological (D-N) explanations. Only later did Hempel (1962) explicitly treat in detail inductive-statistical (I-S) explanations, in which the relevant laws are probabilistic rather than deterministic. All this work led to a fruitful program of research and philosophical debate for many decades. Hempel summarized his analysis of explanation, replying to criticisms and grappling with problems, in his lengthy and detailed 'Aspects of Scientific Explanation' (1965).[13]

As originally presented in 1948, the D-N model arises from adopting four conditions of adequacy for a genuine explanation: three logical, one empirical. The logical (or formal) conditions are that (1) the explanandum (the sentence stating what stands in need of explanation) must be logically deducible from the explanans (the sentences giving the explanation), (2) the explanans must contain at least one general law and (3) the explanans must have empirical content.[14] The single empirical (or material) condition is that the sentences in the explanans must be true.

The D-N model raises difficult philosophical problems about laws which Hempel confessed that he could not solve satisfactorily. One concerns the characterization of laws in strictly logical or formal terms. Following the logical empiricist tradition, laws are represented as universal generalizations (e.g., all copper conducts electricity), but not all true universal generalizations are laws (e.g., all the coins in my pocket are dimes). As Nelson Goodman and others pointed out, laws permit counterfactual inferences that merely accidental true generalizations do not. (Compare 'If this glass rod were made of copper, then it would conduct electricity' with 'If this penny were in my pocket, then it would be a dime.') Hempel recognized that merely requiring that laws be expressed using general predicates (for example, terms that make no reference to any particular objects, spatial regions or time periods and that could apply to an unlimited number of things) would not work. In *Philosophy of Natural Science* (1966), Hempel followed Ernest Nagel by taking the pragmatic route: a universal statement counts as a law if it has a basis in an empirically well-confirmed scientific theory (and does not rule out anything that other accepted theories imply is possible).

Another problem with laws concerns their explanation. Though studies of explanation in the Hempelian tradition have tended to focus on the explanation of particular events and singular facts, Hempel aspired to include the explanation of laws within his model. His conditions of adequacy were deliberately crafted so as to allow for the explanation of laws by deducing them from other laws that are more general in scope, thus reflecting a common scientific pattern. The problem that arises, and which Hempel admitted that he was unable to solve, is the so-called conjunction problem. Suppose that we conjoin two unrelated laws, say Kepler's first law of planetary motion (K) and Boyle's law of gases (B). From the conjunction, (K & B), we can trivially derive B, but obviously this would not count as any sort of explanation of Boyle's law.[15]

What was the rationale for the covering law model that Hempel (and other logical empiricists, as well as Popper) found so compelling?

In part, the model was naturally suggested by the way in which work-ing scientists explain why a particular chemical reaction occurred or why a planet moves in an elliptical orbit, by deducing that behaviour from relevant laws. In some areas at least (especially physics and chemistry) it matched scientific practice.[16] On the philosophical side the model was appealing because (1) it was rigorous, (2) it was objective, (3) it made explanations empirically testable and (4) it avoided appeals to murky ('metaphysical') notions, such as causation. Though Hempel denied that explanation should be restricted to causal explanation, he, like his fel-low empiricists, followed David Hume in insisting that the empirical, testable content of any causal claim (e.g., *A* is caused by *B*) is contained in a universal generalization (All *A*s are followed by *B*s). For empiricists, the covering law model captured what was philosophically respectable in the notion of causal explanation while also allowing a precise explica-tion of that broader class of non-causal scientific explanations involving laws. Vague appeals to 'entelechies', or vital forces, in biology, for exam-ple, were immediately dismissed as non-explanatory since they were not accompanied by any testable laws according to which the alleged entities operate.

Hempel always insisted that two of the main goals of scientific theorizing are explanation and prediction and that these are distinct. Science is not just about predicting the outcomes of observation and experiment; it is also vitally concerned with explaining and understand-ing the world. Nonetheless, the covering law model implies that expla-nation and the prediction of particular events are formally identical. In a series of papers, culminating in 'Aspects', Hempel defended the so-called symmetry thesis, according to which every adequate explanation is potentially a prediction and every adequate prediction is potentially an explanation. The difference between them is pragmatic; it depends on what information was available at the time that the explanatory argument or the predictive argument was given and the purpose for which the argument is being made. Explanations are sought for events that we already know have occurred; predictions concern events that we have not yet observed. With great ingenuity Hempel responded to alleged counterexamples to the symmetry thesis proposed by Michael Scriven, Israel Scheffler and others. Some of these cases involved our alleged ability to explain events after they have occurred even though we would have been hard pressed to predict them ahead of time. Other cases describe arguments that do not explain what they successfully predict. For example, we can calculate the height of a flagpole from the length of its shadow using the laws of optics and simple geometry, but it is hard to accept that this derivation (which satisfies the D-N model) explains why the flagpole has that particular height. In 'Aspects', Hempel acknowledged that this and similar examples give some grounds for doubting at least part of the symmetry thesis: not every adequate prediction is potentially an explanation.

Perhaps the most telling objection to the D-N model is that, given what Hempel and the logical empiricists in general accept as laws, namely, true universal generalizations, nomic (lawlike) derivability fails to capture the notion of explanatory relevance. Wesley Salmon and Peter Achinstein, among others, pointed out that all men who take birth-control pills fail to become pregnant and that all 'hexed' salt (that is, salt over which a spell has been cast) dissolves in water. Yet we do not accept derivation from these laws as explaining why a particular man who ingests birth-control pills fails to conceive or why a particular sample of hexed salt dissolves in water. Hempel's reply to this objection was uncompromising: as long as the conditions of the D-N model are met, we have a fully adequate explanation. The irrelevant information contained in the explanans does not spoil the explanatory value of the argument. Despite Hempel's stand, most philosophers who now reject the D-N model regard the relevance objection as decisive. Many see as its underlying weakness its failure to pick out the factors that are *causally* relevant to the explanandum.

Like his work on the D-N model, Hempel's development of the I-S model of explanation was skilful, innovative and influential. The technical issues involved cannot be discussed here, but, as with the D-N model, there are some general problems with Hempel's approach to probabilistic explanation that many philosophers of science have found telling. An I-S explanation, on Hempel's account, is an instance of the inductive argument type statistical syllogism. The premise (explanans) have the form $Pr(G/F) = r$, and Fa; the conclusion (explanandum) is Ga. Among the logical conditions of adequacy, Hempel requires that the explanandum follows from the explanans with high inductive probability. This means that for an I-S explanation to be adequate, r, in the statistical law $Pr(G/F) = r$, must be close to 1. To many critics of Hempel's approach this is unacceptable since it would make it impossible to explain improbable events. The error, in their view, is to conflate the strength of a probabilistic explanation (how good it is) with the degree to which the explanandum event was to be expected given the information in the explanans. Here is an example (from Peter Railton 1978) that makes the point in a vivid way. Suppose we have a genuinely random wheel of fortune with 99 red stops and 1 black stop. Since the wheel is genuinely random, once it has begun to spin, nothing can affect where it comes to rest. Each of the 100 stops has exactly the same chance of being chosen. Under these circumstances, it seems absurd to insist that we can explain why the wheel halts at a red stop but not why it halts at the black one. Surely, the explanation is equally good in either case, since it is based on an irreducibly indeterministic mechanism and, by hypothesis, our explanatory information is complete. Admittedly the prediction that the wheel will halt on red is more secure than the prediction that the wheel will halt on black. But we should not confuse the strength of the predictive inferences we can make from a set of premises with their explanatory power. In response to this kind of objection (from Salmon, Jeffrey and others), Hempel finally dropped the high-probability requirement in his postscript to the German edition of *Aspects* (Hempel 1977).

Another controversial feature of the I-S model stems from Hempel's empirical conditions of adequacy. Just as with the D-N model, the I-S model requires that the sentences in the explanans must be true; but Hempel imposed a further empirical condition on I-S explanations, namely, the requirement of maximal specificity (RMS). The motivation for the RMS is to solve what Hempel calls the problem of the ambiguity of statistical explanation. There might be (and often are) several different statistical generalizations under which we could place an event in order to explain it. All these generalizations (statistical laws) are true, but they assign different values of the probability r, depending on the reference class to which the event Ga is referred. Which one should we choose? Even worse, from Hempel's point of view, is the possibility that our 'body of knowledge, K', might contain statistical laws $Pr(G/F) = r$ and $Pr(\sim G/H) = r$ (which both satisfy the high-probability requirement) so that we could explain Ga (when a turns out to be G) and also give an equally strong explanation of $\sim Ga$ (should a not turn out to be G). Hempel found this to be intolerable because it would allow us to use K to predict both Ga and $\sim Ga$, thus generating what he called inductive inconsistencies.

As Hempel wrestled with this problem, his proposals went through several refinements in response to criticisms, but all his attempted solutions had a common feature: they made the RMS relative to the beliefs accepted by scientists at a given time.[17] Hempel calls this feature of I-S explanations their epistemic relativity. It means that, unlike D-N explanations (for which the only empirical condition of adequacy is the truth of the explanans), there is no such thing as an objective, 'correct' I-S explanation independent of the scientific context. Hempel's critics found this unacceptable. Their complaint is not that which explanations we accept as correct at a given time depends on our beliefs. What is deemed to be incoherent is the claim that what it is to be a correct explanation depends on our beliefs. This would be analogous to claiming that what it is for a proposition to be true depends on our beliefs. In either case, it is only if being a correct explanation and being true are belief-independent that we can make sense of judging of a particular proposition that is true or of a particular explanation that it is correct.[18]

The structure of theories

Like his mentor, Reichenbach, Hempel was always favourably disposed to realism about scientific theories and the entities and processes they postulate. He argued strenuously against instrumentalism, that is, the view that theories are merely convenient instruments for inferring observation statements from other such statements. Very crudely, the instrumentalist sees that $(T \& O_1) \supset O_2$ and then argues that we could dispense with T and replace it with $(O_1 \supset O_2)$. In 'The Theoretician's Dilemma' (1958) Hempel showed the limitations of results in formal logic (such as Craig's theorem) for replacing theories with sets of statements that use only observation terms. The basic flaw in instrumentalism is that it emphasizes just one aspect of theories – their predictive ability – while ignoring their explanatory power and the goal of achieving a systematic, unified view of the world.

By the 1950s, an account of the structure of scientific theories had emerged, thanks to the work of Carnap, Hempel and other logical empiricists, that was dubbed the 'received,' 'standard' or 'orthodox' view (Suppe 1977). According to the received view, science has two layers: the theoretical and the observational. The observational layer (O)

has two levels: observation statements about particular events and facts at the bottom, with experimental laws (represented as universal generalizations) above them. Everything in these two levels is expressed solely using observational terms whose meaning is accepted as uncontroversial. Observation statements can be quickly verified or confirmed. Experimental laws (expressed solely in observational language) can be inductively confirmed by observation statements; and those laws can be used to predict and explain particular events.

The top layer contains scientific theories. These consist of theoretical postulates (*T*) and correspondence rules (*C*). The theoretical postulates are the axioms of the theory expressed solely in theoretical terms using first-order predicate calculus. The theoretical terms (represented by letters and variables) in the axioms are regarded as uninterpreted symbols: by themselves, theoretical terms have no meaning apart from the formal logical relations between them. The correspondence rules (or what Carnap called meaning postulates) are sentences that connect the theoretical and observational layers. Each correspondence rule (*C*-rule) is a mixed sentence, including both theoretical terms and observational terms. It is the *C*-rules that forge the logical link between the theoretical postulates at the top and the observational statements and empirical laws at the bottom. In a highly schematic representation, we have

$$(T \,\&\, C) \supset O$$

The 'theory' is the conjunction of *T* and *C*. The *C*-rules do double duty on the logical empiricist picture. They enable the theoretical postulates to be tested and confirmed; they also, in Herbert Feigl's vivid phrase, enable 'an "upward seepage" of meaning from the observational terms to the theoretical concepts' (Feigl 1970, 7), thus conferring on the theoretical terms whatever empirical significance they come to have. *C*-rules are instruments of indirect confirmation and, thereby, of partial empirical interpretation.

In his later writings Hempel (1970, 1973) abandoned the assumption (from his Vienna days) that there is a single, canonical observation language. Rather, he insisted that the boundary between theoretical concepts and observational ones is fluid and changes with scientific progress. The theoretical terms of earlier science often become part

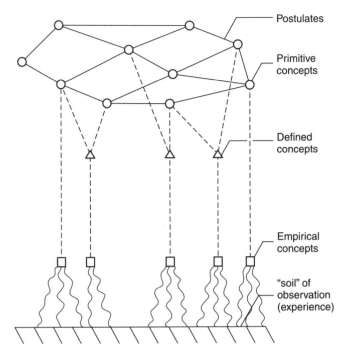

Figure 4.1 Herbert Feigl's diagram of the 'standard' logical empiricist view of theories. (Reproduced from Feigl 1970 with the kind permission of University of Minnesota Press.)

of the observational vocabulary of later science. In this way, Hempel was reflecting the views of Kuhn, Hanson and Feyerabend, all of whom emphasized that observation reports in science are laden with theoretical assumptions. Hempel came to regard the assumption that the interpretation of theoretical terms in scientific theories rest on a strictly observational base as unnecessarily narrow and artificial. Rather, he appealed to 'antecedently available vocabulary' as the means by which scientists understand new theoretical terminology and decide which phenomena theories should explain and be tested against. There is no privileged class of sentences (such as C-rules) that define the meaning of each theoretical term. In appealing to agreement among scientists in a given field at a particular time as determining what counts as the relevant evidence and the vocabulary in which that evidence is described,

Hempel explicitly recognized a historical and pragmatic dimension to the analysis of scientific theories.

Of Hempel's late papers, his work on 'the problem of provisos' has provoked most philosophical attention and marks a further shift away from the received view. Hempel (1988) argued for the limitations of deductive logic as an account of scientific inference. He gives a simple example from the theory of magnetism in which the theoretical postulates include the laws of magnetic attraction and repulsion expressed in terms such as *magnet, north pole* and *south pole* for which interpretative sentences give operational criteria. For example, if iron filings cling to a metal bar, then the bar is a magnet. The theoretical postulates include the principle that if a bar magnet is snapped in half, the two parts thus created will also be magnets whose unlike poles attract and like poles repel. Testing this principle involves further interpretative sentences that link the attraction and repulsion of magnets with their observable behaviour when suspended close to one another by long, thin threads.

Hempel makes two points, the first of which is uncontroversial. The transition from the datum sentence 'iron filings cling to this metal bar' to the theoretical hypothesis 'this metal bar is a magnet' is not deductively valid. The bar could be made of lead and the iron filings attracted by a magnet hidden beneath it; or the bar could be iron but unmagnetized and covered with glue. It is widely recognized that this kind of theoretical ascent is not deductive but inductive (possibly an inference to the best explanation). Second, and more important, the inference in the reverse direction, from theoretical postulates to testable consequences, also fails to be deductive. It is this feature of 'theoretical descent' that Hempel calls the problem of provisos.[19]

Provisos are those presuppositions, usually unstated, that are essential to particular applications of scientific theories.[20] In the magnet example, we presuppose that disturbing forces are absent when we predict the alignment of suspended metal bars from the theoretical hypothesis that both bars are magnets. Those disturbing forces might be due to air currents, electric fields, or magnetic fields. Ascertaining that such forces are absent would require using scientific theories beyond the simple theory of magnetism, and the application of those additional theories will in turn depend on further provisos.

From the inevitability and nature of provisos, Hempel draws a number of morals. First, interpretative sentences alone cannot provide complete or even partial criteria for applying theoretical terms.[21] Second, the ambiguity of falsification extends beyond the form envisaged by the Duhem-Quine thesis since provisos are involved no matter how many additional hypotheses are added to one's original theory in order to generate a testable consequence. Third, the problem of provisos blocks formal attempts (such as those based on Ramsey sentences or Craigian reduction sentences) to replace scientific theories with observational equivalents. More generally it implies the inadequacy of a purely instru-mentalist conception of theories. If theories cannot function merely as devices for getting from some data sentences to others, this sup-ports a realist interpretation of scientific theories and the entities they postulate.

Hempel is quite clear that he does not regard provisos as a vague, all-purpose, general pronouncement attached to every theory proclaim-ing that 'everything else is equal.' Rather, provisos are specific to each application of a theory, certifying that in a particular situation, all the relevant factors have been taken into account. Deciding which factors are relevant is a matter of scientific judgment, learned during one's professional training and career. Hempel agrees with Kuhn that this is another area of scientific reasoning that cannot be given a precise expli-cation or reduced to unambiguous rules.

Kuhn and the rationality of science

Hempel and Kuhn agree that science is a rational enterprise and indeed the best means we have for gaining reliable knowledge about the world. The problem is whether Kuhn is entitled to this judgment given his view of scientific change presented in *The Structure of Scientific Revolutions.* For it seemed to many critics that, in emphasizing the limitations of rea-soned argument, evidence and shared methodological standards during periods of 'revolutionary science,' Kuhn had in effect (in the words of Imre Lakatos) reduced scientific change to 'mob psychology.' Kuhn vehemently denied the charge of irrationalism and insisted that all scientists share core epistemic standards for assessing scientific theories, even at the paradigm level. Among the most important of these 'values' (in Kuhn's terminology) are accuracy, consistency, scope, simplicity and fruitfulness. Though scien-tists will often disagree about how to interpret these standards and how to rank them, scientists' commitment to these shared values suffices to make scientific debate rational, even though the outcome of that debate cannot be captured by any algorithm or system of inductive logic.

Though he did not dispute Kuhn's historical description of scientific revolutions and found much of value in Kuhn's work, Hempel doubted that Kuhn (and the 'historic-sociological' or 'pragmatist' school inspired by Kuhn) had shown in a satisfactory manner that science is rational.[22] Furthermore, Hempel insisted that the analytic empiricist approach remains valuable precisely because it addresses the normative, epistemic dimension of science that Kuhn's naturalistic view fails to capture. In Hempel's view, Kuhn did not appreciate adequately the distinction (insisted on by Reichenbach, Carnap and Popper) between describing how someone in fact reasons and justifying standards that govern how someone ought to reason. The first is an empirical question of psychology (or sociology or history); the second is a normative question of logic and epistemology.[23]

Throughout his career Hempel advocated an instrumentalist conception of rationality according to which to say that someone acted rationally implies that the agent had a certain goal and adopted means that are suited to achieving that goal. Therefore, to exhibit the rationality of a choice between rival scientific theories requires identifying the goal (or goals) of science and then showing that the methodological principles adopted by scientists when they debate rival theories conduce to achieving that goal (or goals). Even if scientists share certain values that influence their decisions – Hempel calls them desiderata – that by itself does not exhibit the rationality of their choices. Hempel surmises that Kuhn may have been misled by the ambiguity of the phrase 'accounting for scientific choices' into conflating a descriptive explanation of scientific behaviour with a normative justification of the principles on which it is based.[24]

Kuhn is sceptical that the analytic approach called for by Hempel is likely to succeed, especially given the very modest progress made by philosophers of science in solving the problem of induction or constructing a theory of confirmation theory adequate for anything beyond highly idealized formal models. Moreover, Kuhn thinks that such philosophical endeavours are unnecessary since we already have a paradigm of rationality, namely, science itself. What better way to figure out how to acquire reliable knowledge of the world than to study what criteria scientists actually employ in judging rival theories. Using criteria such as accuracy, scope and simplicity is constitutive of science; the aim of science is to judge theories according to the criteria shared by practicing scientists.

Hempel finds Kuhn's gambit of defining science in its own terms an evasion of the demand for an 'aim/means' analysis of scientific rationality. He concedes that the analytic approach has made scant progress so far in constructing adequate models of explanation, confirmation and the like that can do justice to the rich detail of scientific practice; the individual epistemic values of science have not been explicated with the kind of precision that would permit competing theories to be compared, and even if such precision had been achieved, there is no guarantee that the various epistemic standards could be combined so as to yield a single numerical score for each theory. Nonetheless, this does not entail that these efforts cannot succeed, perhaps at a level less global than that of Kuhnian paradigms. Understandably, up until now philosophers have worked on idealized models that go some way towards capturing the underlying logic of small-scale science. Most analytic empiricists have not attempted to tackle the big revolutionary changes in science on which Kuhn focused his attention.[25] But these preliminary steps, modest though they are, have been shaped by the desire to reveal and justify scientific reasoning, and without such an understanding, it is difficult to explain why our confidence in our best science is rationally justified.

One matter on which Hempel thinks Kuhn has been judged unfairly is the accusation of mob psychology. To sustain a charge of irrationality would require showing that according to Kuhn, scientists violate or ignore accepted standards of rationality when paradigms change. As far as Hempel can tell, Kuhn does not claim this. Epistemic values are imprecise and various, they are apparently irreducible one to another and comparative rankings among them can be difficult to adjudicate. But there is not obviously anything missing from Kuhn's list of desiderata, and Kuhn explicitly denies that scientists would ever deliberately favour the weaker theory if the contest between two rivals had a single dimension (such as, say, accuracy, all other factors being equal).

In his final decade, Hempel published a number of papers taking stock of his career. He placed logical empiricism in historical perspective and offered a much more favourable assessment of Kuhn's historico-sociological turn in philosophy of science.[26] These papers (conveniently collected in Hempel 2000) emphasize the pragmatic dimension of philosophy of science, which Hempel traced back to 'the other Vienna Circle' of Otto Neurath (Wolters 2003). Philosophers of science are urged to pay close attention to the realities of scientific practice before attempting to construct normative methodologies. The traditional logical-empiricist goal of using only logic and the analysis of concepts to arrive at a timeless, rational account of science (in the manner of Popper, Carnap and the earlier Hempel) is dismissed as a chimera.

Notes

The author would like to thank Jan Cover, Chris Pincock, and Patricia Curd for their valuable help and suggestions.

1 Thomas S. Kuhn, *The Road Since Structure*, ed. by James Conant and John Haugeland (Chicago: University of Chicago Press, 2000), p. 226.

2 The Berlin circle also included Paul Oppenheim, Kurt Grelling, Walter Dubislav and Olaf Helmer, with whom Hempel had several fruitful collaborations. See Rescher (2006) for the distinctive character of the Berlin version of logical empiricism.

3 Soon after Hitler came to power in 1933, the Nazis built one of their first concentration camps in Oranienburg, replaced two years later by the notorious Sachsenhausen extermination camp. Kurt Grelling and his wife died in the gas chambers of Auschwitz; Janina Lindenbaum-Hossiasson (a Polish logician who influenced Hempel's early work on confirmation) was murdered by the Gestapo.

4 Hempel credited Oppenheim with having saved his life twice: once when he enabled Hempel to leave Nazi Germany and again when Oppenheim insisted that Hempel be treated following a life-threatening injury in a bicycle accident.

5 Originally founded in 1930 by Carnap and Reichenbach, *Erkenntnis* had ceased publication in 1939 at the outbreak of World War II.

6 Though it appeared as the final entry in the *International Encyclopedia of Unified Science* – a series that had been founded by Neurath, Carnap, and Charles Morris in 1937 to promote Neurath's vision of the 'unity of science' and the views of the Vienna Circle – Kuhn's book was regarded by some philosophers as antithetical to logical positivism and logical empiricism. For an influential revisionary account of the relationship between Kuhn's work and logical empiricism and its cordial reception by Carnap and others, see George Reisch, 'Did Kuhn Kill Logical Empiricism?' *Philosophy of Science* 58 (1991): 264–77. In response to Reisch and others, J. C. Pinto de Oliveira, 'Carnap, Kuhn, and Revisionism: On the Publication of *Structure* in *Encyclopedia*', *Journal for General Philosophy of Science* 38 (2007): 147–57 points out that Charles Morris (associate editor with Carnap of the *Encyclopedia*) had commissioned Kuhn's

work as the entry on the history of science, and for that reason alone Carnap would not have found it inconsistent with his own philosophical approach to understanding science.

7 Schlick retained the complete verifiability principle by arguing that scientific laws are not statements (and hence neither true nor false) but rules or instruments for inferring some statements from others.

8 S appears vacuously in a compound sentence N if N is logically equivalent to another sentence in which S does not appear. A simple example would be if the only appearance of S in N was in the form of the tautology $(S \lor \sim S)$.

9 Hempel's adequacy condition has been questioned by David Rynin, Wesley Salmon and others. See Salmon (1966).

10 We could blame this problem on the material conditional and insist that what we need to do justice to disposition terms are counterfactual conditionals, which are subjunctive and non-truth-functional. For example, an object is fragile if and only if it *would* break if it *were* sharply struck, a fact that is true even if the object is never actually struck. But how are we to analyse counterfactual conditionals in terms of observables? One popular approach, proposed by Nelson Goodman, analyses them in terms of laws of nature. This raises further problems for empiricists since laws of nature are not themselves observable or obviously reducible to observables.

11 The quantitative work, performed in collaboration with Olaf Helmer and Paul Oppenheim, was in friendly rivalry with Carnap's program to establish a system of inductive logic. Carnap referred to the Hempel-Helmer-Oppenheim approach as the H_2O theory. See Rescher (1997).

12 Another feature of Hempel's account that many modern confirmation theorists reject (especially Bayesians) is Hempel's insistence that if E confirms H, then E must also confirm anything that is entailed by H. See Fitelson (2006).

13 For detailed surveys of work on explanation, including the contributions of Hempel, see Salmon (1989) and Psillos (2007).

14 As Hempel points out, condition (3) is not strictly needed since it follows from (1); anything that implies an explanandum that has empirical content must itself have empirical content.

15 This conjunction problem has since been addressed by Philip Kitcher's unification model of explanation, in which, as in Hempel's account, explanations are conceived as special kinds of inference or argument. See Kitcher (1981).

16 Things are apparently different in some areas of biology, history, and anthropology in which appeal to the fact that some property P of X is the function of X is deemed to explain why X has P, even though that function could be performed by some other property of X. Because they do not entail what they purport to explain, Hempel denied that functional ascriptions are explanatory.

17 Though Hempel uses the phrase 'knowledge situation, K', he means the set of sentences that scientists accept as true. Thus, K almost certainly contains some false sentences, and the contents of K will change with time.

18 See Railton (1978) and Coffa (1974). Hempel (1977) acknowledges the force of these objections to his thesis that I-S explanations and statistical laws are essentially relative to our knowledge situation. He concedes that an absolute (non-epistemically relative) explication of these notions is both possible and desirable.

19 Hempel also thinks provisos are crucially involved in theoretical ascent.

20 Hempel points out that provisos are also involved at the theoretical level in inferring one theoretical claim to another. For example, in inferring that a magnet produces other magnets when broken in pieces, we presuppose that the temperature is not too high.

21 Hempel rejects the possibility that provisos might be expressed as interpretative sentences of probabilistic form since any attempt to specify the relevant probabilities (regarded as empirical frequencies) would involve considerations that go far beyond the original theory.

22 See Hempel, 'Scientific Rationality: Normative vs. Descriptive Construals' (1979), and Hempel, 'Valuation and Objectivity in Science' (1983).

23 Kuhn was certainly aware of the distinction. In the introduction to *The Structure of Scientific Revolutions* he writes, 'In the preceding paragraph I may even seem to have violated the very influential contemporary distinction between "the context of discovery" and "the context of justification"'. He then goes to ask, rhetorically, 'Can anything more than profound confusion be indicated by this admixture of diverse fields and concerns?' (*Structure*, pp. 8–9).

24 It should be noted that this dispute between Kuhn and Hempel is not the familiar disagreement between internalists and externalists about how to understand scientific change. Though sociologists of science often claim Kuhn as an inspiration for their approach, Kuhn himself remained resolutely internalist with regard to explaining nearly all aspects of scientific evolution, including scientific revolutions.

25 One notable exception is Karl Popper; another is Wesley Salmon, who sought to reconcile Kuhn with logical empiricism by using Bayes's theorem to analyse comparative theory choice in science. Salmon allows Kuhn's 'values' to affect the prior probabilities of theories and thus influence the confirmation they receive from evidence. Hempel was doubtful that Salmon's proposal could work. See the papers by Kuhn, Salmon and Hempel on rationality and theory choice in *The Journal of Philosophy* 80 (1983): 555–72.

26 See especially 'The Status and Rationale of Scientific Methodology' (1998) and 'The Irrelevance of Truth for the Critical Appraisal of Scientific Theories' (1990). In the latter, Hempel argues that science aims not at truth but at achieving world pictures that are 'epistemically optimal', something that changes with time and is relative to scientists' beliefs. While the concept of truth is indispensable for logic and semantics, scientific choices depend on applying Kuhn's desiderata (scope, simplicity and the like), methodological factors, often vague, that have no provable connection with truth.

References

Coffa, J. Alberto (1974), 'Hempel's Ambiguity'. *Synthese*, 141–63.

Feigl, Herbert (1970), 'The "Orthodox" View of Theories: Remarks in Defense as well as Critique', in M. Radner and S. Winokur (eds), *Minnesota Studies in the Philosophy of Science*, vol. IV. Minneapolis: University of Minnesota Press, 3–40.

Fetzer, James H. (ed.) (2000), *Science, Explanation, and Rationality: Aspects of the Philosophy of Carl G. Hempel*. New York: Oxford University Press.

Fitelson, Branden (2006), 'The Paradox of Confirmation'. *Philosophy Compass* 1, 95–113.

Friedman, Michael (2000), 'Hempel and the Vienna Circle', in Fetzer (2000), 39–64; reprinted in Hardcastle and Richardson (2003).

Hardcastle, Gary L. and Alan W. Richardson (eds) (2003), *Minnesota Studies in the Philosophy of Science, Volume XVIII, Logical Empiricism in North America*, Minnesota: University of Minnesota Press.

Hempel, Carl G. (1945), 'Studies in the Logic of Confirmation'. *Mind*, 54, 1–26 and 97–121; reprinted in *Aspects* (1965) with a Postscript.

— (with P. Oppenheim) (1948), 'Studies in the Logic of Explanation'. *Philosophy of Science*, 15; reprinted in *Aspects* (1965) with a Postscript.

— (1950), 'Problems and Changes in the Empiricist Criterion of Meaning'. *Revue Internationale de Philosophie*, No. 11, 41–63.

— (1951),'The Concept of Cognitive Significance: A Reconsideration'. *Proceedings of the American Academy of Arts and Sciences*, 80, 61–77.

— (1952), *Fundamentals of Concept Formation in Empirical Science*, vol. II, no. 7, *International Encyclopedia of Unified Science*. Chicago: University of Chicago Press.

— (1954), 'A Logical Appraisal of Operationism'. *Scientific Monthly*, 79, 215–20; reprinted in *Aspects* (1965).

— (1958), 'The Theoretician's Dilemma', in H. Feigl, M. Scriven, and G. Maxwell (eds), *Minnesota Studies in the Philosophy of Science*, vol. II. Minneapolis: University of Minnesota Press, 37–98; reprinted in *Aspects* (1965).

— (1962), 'Deductive-Nomological vs. Statistical Explanation', in H. Feigl and G. Maxwell (eds), *Minnesota Studies in the Philosophy of Science*, vol. III, (Minneapolis: University of Minnesota Press, 98–169; reprinted in Hempel (2001).

— (1965), 'Aspects of Scientific Explanation', in C. G. Hempel, *Aspects of Scientific Explanation and Other Essays in the Philosophy of Science*. New York: Free Press.

— (1965), *Aspects of Scientific Explanation and Other Essays in the Philosophy of Science*. New York: Free Press, 1965). [Cited in text as '*Aspects*'.]

— (1965), 'Empiricist Criteria of Cognitive Significance: Problems and Changes', in *Aspects*. This article combines Hempel's papers of 1950 and 1951 and adds a brief Postscript.

— (1966), *Philosophy of Natural Science*. Englewood Cliffs, N J: Prentice-Hall.

— (1970), 'On the "Standard Conception" of Scientific Theories', in M. Radner and S. Winokur (eds), *Minnesota Studies in the Philosophy of Science*, vol. IV. Minneapolis: University of Minnesota Press, 142–63; reprinted in Hempel (2001).

— (1973), 'The Meaning of Theoretical Terms: A Critique of the Standard Empiricist Construal', in P. Suppes et al. (eds), *Logic, Methodology and Philosophy of Science IV*. Amsterdam: North Holland Publishing Company, 367–78; reprinted in Hempel (2001).

— (1977), 'Postscript 1976: More Recent Ideas on the Problem of Statistical Explanation', in C. G. Hempel, *Aspkete wissenschaftlicher Erklärung*. Berlin and New York: Walter de Guyter; trans. by Hazel Maxian in Hempel (2001).

— (1979), 'Scientific Rationality: Normative vs. Descriptive Construals,' in H. Berghel, A. Huebner, E. Koehler (eds), *Wittgenstein, the Vienna Circle, and Critical Rationalism*. Proceedings of the Third International Wittgenstein Symposium, August 1978. Vienna: Hoelder-Pichler-Tempsky, 291–301; reprinted in Hempel (2001).

— (1983), 'Valuation and Objectivity in Science', in R. S. Cohen and L. Laudan (eds), *Physics, Philosophy and Psychoanalysis: Essays in Honor of Adolf Grünbaum*. Dordrecht: D. Reidel, 73–100; reprinted in Hempel (2001).

— (1988), 'Limits of a Deductive Construal of the Function of Scientific Theories', in E. Ullmann-Margalit (ed.), *Science in Reflection. The Israel Colloquium*, vol. 3. Dordrecht: Kluwer Academic Publishers, 1–15; reprinted in Hempel (2001).

— (1988), 'On the Cognitive Status and the Rationale of Scientific Methodology'. *Poetics Today*, 9, 5–27; reprinted in Hempel (2000).

— (1988), 'Provisoes: A Problem Concerning the Inferential Function of Scientific Theories'. *Erkenntnis*, 28, 147–64; reprinted in Hempel (2000).

— (1990), 'Il significato del concetto di verità per la valutazione critica delle teorie scientifiche'. *Nuova Civiltà delle Macchine*, Anno VIII, N. 4, 32, 7–12; reprinted as 'The Irrelevance of Truth for the Critical Appraisal of Scientific Theories' in Hempel (2000).

— (2000), *Selected Philosophical Essays: Carl G. Hempel*, Richard Jeffrey (ed.). Cambridge: Cambridge University Press.

— (2001), *The Philosophy of Carl G. Hempel: Studies in Science, Explanation, and Rationality*, James H. Fetzer (ed.). New York: Oxford University Press.

Kitcher, Philip (1981), 'Explanatory Unification'. *Philosophy of Science*, 48, 507–31.

Psillos, Stathis (2007), 'Past and Contemporary Perspectives on Explanation', in Theo A. F. Kuipers (ed.), *Handbook of the Philosophy of Science, General Philosophy of Science: Focal Issues*. Amsterdam: Elsevier, 97–174.

Railton, Peter (1978), 'A Deductive-Nomological Model of Probabilistic Explanation'. *Philosophy of Science*, 45, 200–26.

Rescher, Nicholas (1997), 'H_2O: Hempel-Helmer-Oppenheim, An Episode in the History of Scientific Philosophy in the Twentieth Century'. *Philosophy of Science*, 64, 334–60.

— (2006), 'The Berlin School of Logical Empiricism and Its Legacy'. *Erkenntnis*, 64, 281–304.

Rosenkrantz, Roger D. (1982), 'Does the Philosophy of Induction Rest on a Mistake?,' *Journal of Philosophy*, 79, 78–97.

Salmon, Wesley C. (1966), 'Verifiability and Logic', in P. K. Feyerabend and G. Maxwell (eds), *Mind, Matter, and Method: Essays in Honor of Herbert Feigl*. Minneapolis: University of Minnesota Press, 354–76.

— (1989), 'Four Decades of Scientific Explanation', in P. Kitcher and W. C. Salmon (eds), *Minnesota Studies in the Philosophy of Science*, vol. 13. Minneapolis: University of Minnesota Press, 3–219.

— (1999), 'The Spirit of Logical Empiricism; Carl G. Hempel's Role in Twentieth-Century Philosophy of Science'. *Philosophy of Science*, 99, 333–50; reprinted in Fetzer (2000) 309–24.

Scheffler, Israel (1963), *The Anatomy of Inquiry: Philosophical Studies in the Theory of Science*. New York: Knopf.

Suppe, Frederick (1977), 'The Search for Philosophical Understanding of Scientific Theories', in F. Suppe (ed.), *The Structure of Scientific Theories*, 2nd edn. Urbana: University of Illinois Press.

Wolters, Gereon (2003), 'Carl Gustav Hempel: Pragmatic Empiricist', in Paolo Parrini, Wesley C. Salmon and Merrilee H. Salmon (eds), *Logical Empiricism: Historical and Contemporary Perspectives*. Pittsburgh: University of Pittsburgh Press, 109–22.

CHAPTER 5

ANTI-INDUCTIVISM AS WORLDVIEW: THE PHILOSOPHY OF KARL POPPER

Steve Fuller

Introduction: the problem of positioning Popper

If philosophers are judged by their entire corpus, Karl Popper was arguably the most noteworthy if not the greatest philosopher of the twentieth century. It is difficult to imagine a field of academic or public life that his work did not touch directly or indirectly. (If in doubt, consult Shearmur and Norris 2008, which includes Popper's media interventions.) Indeed, Popperian buzzwords still populate the trading zone between academic and public discourse – 'falsifiability', 'demarcation criteria', 'the open society', 'the poverty of historicism', 'methodological individualism', 'conjectures and refutations', 'evolutionary epistemology', 'world three' (i.e., the realm of objective knowledge). To be sure, in many cases Popper was responsible more for promoting than developing the ideas behind the buzzwords. But most remarkable, especially for a twentieth-century philosopher, Popper's ideas remain memorable in ways that have not proven either embarrassing or shameful.

A disturbing feature of the past century is that the dominant figures of the two main European philosophical traditions – Martin Heidegger and Ludwig Wittgenstein – were decidedly conservative thinkers with strong authoritarian tendencies (Safranski 2000; Gellner 1998). The disturbance is only increased when an avowedly liberal thinker like Richard Rorty (1979) explains the significance of his own hero, John Dewey, in terms of the views that Dewey shared with Wittgenstein and Heidegger, as if Karl Popper had never existed. Had Rorty taken Popper's achievement more seriously, we might have acquired greater immunity to the postmodern predicament, whereby the failure to establish logical foundations for all thought opens the door to an endless proliferation of community-based epistemic standards (Hacohen 2000, 2–3). Indeed, a regrettable sign of our non-Popperian times is that the most natural way to interpret the idea of 'social epistemology' is as a consensus-seeking approach to inquiry based on a division of cognitive labour and trust in expertise – not, as Popper himself did, a set of mutually critical agents whose thoroughly conventionalist approach to disciplinary boundaries invites them to question and reform even fundamental knowledge claims on a regular basis (Jarvie 2001). Of course, some social epistemologists take Popper seriously (Fuller 2000a).

Popper's relative neglect by professional philosophers is arguably more than compensated by the impact of his work on other disciplines, especially the social sciences. In the second half of the twentieth century, Popper stood within academia for the scientific method, objectivity, rationality, liberalism and individualism just as his English patron, Bertrand Russell, had among the public at large. But Popper's direct style, which communicated easily across disciplinary boundaries, has not worn well with philosophical colleagues, who tend to interpret it as an expression of dogmatism. Yet most of Popper's 'positive' views were really negative ones in disguise. His rationalism was tinged by fallibilism that set its sights primarily on the unreflective character of inductive inference. His liberalism was consistently anti-authoritarian to the point of harboring strong reservations about the deference that his friend Friedrich Hayek showed to the market's 'invisible hand' (Hacohen 2000, ch. 10). And Popper's individualism was driven much more by an opposition to the prospect of humanity's absorption into one or another kind of groupthink than any metaphysical adherence to self-interest as the defining feature of human nature. Popper's de facto 'oppositional consciousness' meant that he often presented his views as critical sketches that presumed some acquaintance with the details and history of what was being criticized.

Failure to appreciate the profoundly dialectical character of Popper's thought has led to his portrayal – at the hands of no less than the self-avowed keepers of the dialectical tradition – as a relatively simple-minded thinker, such as the standard-issue 'positivist' that came across to the self-styled 'critical theorists' of the Frankfurt School, Theodor Adorno and Jürgen Habermas, in the *Methodenstreit* of the 1960s (Adorno 1967; cf. Fuller 2003, chs. 13–14). But even among the positivists' Anglo-American descendants in analytic philosophy, in whose ranks he is sometimes misleadingly included, Popper is known for having adopted a distinctive, if not altogether transparent, stance on the role of something called 'induction' in scientific epistemology. Such cagey language is warranted because for Popper the exact nature of induction matters less than the significance it has for him. In essence, 'induction' stands for everything that Popper is against, not only in science but also in politics: *blind conformity to tradition.*

Anti-inductivism as the permanent revolution in the philosophy of science

A good way to get into Popper's understanding of induction is through that old warhorse of analytic epistemology, the so-called grue paradox. According to Nelson Goodman (1955), 'grue' is the property of being green before a given time and blue thereafter. This property enjoys just as much empirical support as the property of being green when hypo- thetically applied to all known emeralds. For Goodman, this was a 'new riddle of induction' because unlike Hume's original example of induc- tion – how do we know that the sun will rise tomorrow just given our past experience? – his problem suggests that our propensity to induc- tive inference is shaped not simply by our prior experience but by the language in which that experience has been cast. Popper could not agree more. Unfortunately Goodman proceeds to draw a conservative conclusion from this situation: namely, that we are generally right to endorse the more familiar predicate 'green' when making predictions about the colour of future emeralds. Why? Well, because that predicate is more 'entrenched', which is a bit of jargon for the rather unphilo- sophical stance of 'if it ain't broke, don't fix it'.

Popper's anti-inductivism may be understood as an attempt to render Goodman's riddle philosophically more interesting. The prospect that a predicate like 'grue' might contribute to a more adequate account of all emeralds (both known and unknown) than 'green' is certainly familiar from the history of science. It trades on the idea that the periodic inability of our best theories to predict the future may rest on our failure to have understood the past all along. In short, we may have thought we lived in one sort of world, when in fact we have been always living in another one. After all, the 'grue' and 'green' worlds have looked exactly the same until now. In this respect, Goodman showed that induction is about locating the actual world in which a prediction is made within the set of possible worlds by proposing causal narratives that purport to connect past and future events, 'green' and 'grue' constituting two alternative accounts vis-à-vis the colour of emeralds.

This is a profound point, especially for scientific realists, the full implications of which have yet to be explored – even now, more than half a century after Goodman's original formulation (cf. Stanford 2006 on the problem of 'unconceived alternatives' to the best scientific explanation at a given time). In particular, seen through Popperian lenses, Goodman suggests how the 'paradigm shifts' that Kuhn (1970) identified with 'scientific revolutions' should be expected if we take the fallibility of our theories as *temporally symmetrical* – that is, that every substantial new discovery is always an invitation to revise what we had believed about the past. In other words, as science increases its breadth by revealing previously unknown phenomena, it increases its depth by revising our understanding of previously known phenomena so as to incorporate them within the newly forged understanding. Thus, Newton did not simply add to Aristotle's project but superseded it altogether by showing that Aristotle had not fully grasped what he thought he had understood. Indeed, if Newton is to be believed, Aristotle literally did not know what he was talking about, since everything that he said that we still deem to be true could be said just as well – and better – by dispensing with his overarching metaphysical framework.

From a Popperian standpoint, science may be defined as the form of organized inquiry that is dedicated to reproducing Goodman's new riddle of induction on a regular basis. To be sure, speculatively conjured predicates like 'grue' rarely lead to successful predictions, so some skill must be involved to make them work, whereby they acquire the leverage for rethinking inquiry's prior history. Such skill in devising 'projectible' predicates, to use Goodman's jargon, is displayed in first-rate scientific theorizing. In this way, scientific revolutions metamorphose from Kuhn's realm of the unwanted and the unintended to Popper's (1981) positive vision of deliberately instigated 'permanent revolutions'.

An unintended consequence of Popper's typecasting as a wayward logical positivist or analytic philosopher is that his historically nuanced interpretation of scientific inquiry has been often overlooked. Consider his distinction in the contexts of scientific discovery and justification (or 'validation', to use Popper's preferred term), the latter making no reference to a knowledge claim's original circumstances (Popper 1959, ch. 1; Fuller 2003, ch. 15). Popper fully recognized both how necessary and how difficult it is to draw the distinction in practice, since even canonical formulations of knowledge claims often bear hints of their origins and aspirations, which can then easily bias evaluation. Although Popper regularly protested the fuss that positivist and analytic philosophers made about semantics, he himself was very alive to the implicit dangers of taking knowledge claims on their face. A seemingly innocent expression like 'Newtonian mechanics' suggests that the theory in question is the legitimate heir of the work of the great late-seventeenth-century physicist, an impression that may serve to inhibit challenges to the theory's foundations.

However, unlike the positivists, Popper did not demand a complete regimentation of the language of science. Rather, he proposed that knowledge claimants specify a 'crucial experiment' – namely, a situation in which two empirically comparable theories predict opposing outcomes under agreed test conditions (Popper 1959, chs. 3–5). Thus, Popper believed in the methodological but not the theoretical unity of science. Put in more philosophically grandiose terms, Popper was an epistemological but not an ontological unificationist. (This point is most developed in Popper 1972.) Popper appeared to believe that we are entitled to our private realities, except when we expect others to abide by them as well. In that case we are obliged to specify what the anthropologist and cybernetician Gregory Bateson (1972), with a nod to the pragmatists Charles Sanders Peirce and William James, dubbed 'the difference that makes a difference'. In other words, what would it take for you to come to believe something that up to this point only I believe? For Bateson this phrase defined what it means to be 'informative' to a given receiver, which is exactly in the spirit of Popper's proposal.

The tricky question for Popperians has been how to understand and institutionalize the outcome of crucial experiments. Francis Bacon, the early-seventeenth-century Lord Chancellor of England who originated the idea, clearly intended it as a peaceful settlement to religious disputes that would otherwise – and eventually did – lead to civil war. Bacon seemed to imagine that the experimental outcomes would become the property of the state, contributing to a common body of knowledge available to govern society in a more rational and less violent way. Indeed, Bacon's proposed House of Solomon is arguably what the Royal Society would have become were it a branch of the civil service rather than a chartered corporation. But otherwise, the larger theoretical frameworks responsible for the competing predictions would be allowed to flourish amongst their self-organized adherents in civil society. From today's standpoint, scientific disputes in this Baconian regime would seem to be very similar to political disputes, with theoretical frameworks functioning as political parties in modern democracies that survive regardless of their record of electoral success. However, there would be a decisive difference. Whereas the tendency in modern democracies has been towards fixed-interval elections, regardless of the achievements or failures of the ruling party, the sort of scientific 'election' implied by crucial experiments for Bacon and Popper is an epistemic 'vote of no confidence' that could be raised by an organized opposition at any time.

Notice what is *not* being affirmed here – namely, the winner-take-all sense of dominance and overriding sense of intellectual purity that is characteristic of a Kuhnian paradigm. For Bacon and Popper, people may form beliefs however they please, but the resulting theoretical vision (or 'metaphysics') becomes a matter of public concern – and hence a candidate for knowledge – only once a formal challenge is made to an opposing theoretical vision. However, the outcome of that challenge is understood as a gain for the public storehouse of knowledge that works at once to enable and constrain subsequent claims to knowledge: All theoretical visions may draw on this knowledge to advance their interests. But equally, if they wish to contest other theoretical visions they must take this knowledge into account when formulating their bids.

We see here the source of Popper's instinctive antipathy to induction: it is rooted in an antipathy to the very idea of a 'legitimate heir' to a well-corroborated scientific theory. For Popper, the whole point of science is that there is no presumption about what counts as an appropriate extension of a theory. In that sense, a theory is no more than a conventionally organized and strategically focused body of evidence aimed at extending inquiry. Here Popper's stance is usefully contrasted with that of W. V. O. Quine, who also recognized that the body of evidence in support of one theory could be equally used to support another theory, even a contradictory one. Quine's considered view of this so-called underdetermination of theory by data was, like Goodman, to plump for a conservative presumption that favours the theory that saves the phenomena with minimal epistemic disruption (Quine and Ullian 1970, ch. 6).

Quine, rather like John Dewey in his epistemological primer for schoolteachers, *How We Think*, was inclined to regard science as technically enhanced natural cognition, an activity focused on 'problem solving' understood in the rather limited sense of biological adaptation, in which an organism optimally 'solves' a problem given by the environment by doing whatever it takes to enable the organism to continue as it has up to that point. In such cases, the 'problem' is whether the current case can be addressed entirely in terms of past experience or requires a different frame of reference. Omitted is the more radical possibility that a different frame of reference is needed for *both* past and present cases. In that case, the problem would be treated less as a localized block than a symptom of some deeper disorder that would precipitate what Kuhn called a 'paradigm shift'. In that case, the subject actively contributes to the construction of the problem space for which she then finds a solution. In effect, by arriving at a new understanding of her past, she opens up a new horizon of epistemic possibilities, the next generation of normal science puzzles. But unlike Kuhn, Popper was more impressed by those who discovered rather than solved problems.

Popper was one with the logical positivists in stressing science's tendency to break with default ways of knowing, even within science itself. In other words, a Kuhnian paradigm shift was precisely when science came into its own as a form of knowledge distinct from the dogma promulgated by propaganda ministries in religious and secular regimes. This attitude reflected the lasting impression that the early-twentieth-century revolutions in relativity and quantum theory left on the intellectual youth of the time. Recall that in 1919, when Einstein's general theory of relativity passed its most widely publicized empirical test (i.e., that light would appear to bend around the sun during a solar eclipse), the positivists were beginning their academic careers and Popper was still a teenager. In effect, they saw the epistemic horizons of the physical universe reconfigured by an empirically successful redefinition of the problem space of physics. Moreover, two of the seminal logical positivists, Rudolf Carnap and Hans Reichenbach, had entered philosophy as refugees from physics just when it presented relatively short-lived but significant resistance to relativity. They went on to promote philosophy as defending a scientific attitude in the face of not only various irrationalist and pseudoscientific tendencies in Weimar culture but also the default conservatism of the scientific establishment that had ostracized them.

Popper's steadfast opposition to presumption in science may be understood partly in terms of this common sensibility that he shared with the positivists. However, he went much further, refusing to see science as itself a 'foundation' on which truth is constructed, a foundation that in turn might be used as the basis for the conduct of public life if not the governance of society more generally. Here it is worth recalling the literalness of the logical positivists' 'positivism'. The Vienna Circle manifesto, 'The Scientific World-View', not only openly acknowledged inspiration from Auguste Comte – the man who aspired to have science exert an authority comparable to that of the Roman Catholic Church in its heyday – but also urged the insertion of vanguard scientific ideas in an envisaged rational reconstruction of post–World War I Europe. These insertions ranged from living spaces to communication systems: that is, from Bauhaus architecture to 'Isotype', a visual Esperanto – or 'universal slang', as Otto Neurath put it (Galison 1990).

In contrast, Popper's adherence to science was always less to its particular first-order theories than its second-order *attitude* towards the world – that is, a rather content-neutral sense of the 'scientific world view' in which experiment functions as a technically enhanced version of Socratic inquiry, as opposed to the positivist impulse to clarify and propagate scientific ideas that have been already secured by technically approved means. A long-term consequence of this subtle but important difference is that Popper and his followers – most notably Paul Feyerabend – often found themselves on the less popular if not losing side of many of the leading scientific controversies of the day, including the mind-body problem, the scientific standing of Darwinian evolution, and especially the consensus – the so-called Copenhagen

Interpretation – that quickly formed around quantum mechanics in the 1920s, which finesses the ontological implications of Heisenberg's uncertainty principle by reducing the goal of physical inquiry to improvement in the prediction of the outcomes of physics experiments. This concordat effectively sealed off from direct scientific consideration wilder yet empirically supported interpretations of what transpired in the key experiments – including radically constructivist views of reality, parallel universes and action at a distance. To be sure, discussion of these theoretical possibilities continued apace but only semi-connected to developments in the empirical side of the physics.

Unlike other philosophers of science who are normally called 'conventionalist', Popper regarded the conventional nature of scientific knowledge claims as a standing challenge rather than a fait accompli. In other words, conventionalism was less interesting as an alternative theory of the epistemic foundations of science than as an indication of the ease with which alternative epistemic foundations could be found for science. Here one might say that Popper and his followers exploited the reflexive implications of conventionalism to turn it into anti-foundationalist theory of epistemic foundations. It is from this standpoint that the philosophy of science known as *instrumentalism* appears as anathema. For Popper instrumentalism is when science loses its existential boldness and slips into 'mere' technology, rendering it intellectually sterile and a pliant tool for the powers that be (Popper 1963, ch. 3).

The spirit of Popper's critique of instrumentalism is worth noting as the mirror image of what Pierre Duhem found attractive in the position. On the one hand, Duhem the committed Catholic followed Galileo's chief inquisitor, Cardinal Robert Bellarmine, in welcoming instrumentalism's humble realization that scientific inquiry could never resolve fundamental differences in theoretical horizons, thereby keeping the door open to faith – hence, Duhem's brand of theism is called 'fideism'. (For an updated version of this sensibility, cf. van Fraassen 1980, esp. ch. 7). On the other, Popper denounced instrumentalism for selling short science's potential, effectively shoring up whatever happens to be the current paradigm. To be sure, the Catholic Church might justify instrumentalism as a piece of pure realpolitik: Social order is best maintained by opportunistically co-opting the dominant theory by allowing it a carefully delineated domain within which to pursue its inquiries, as long as it does not stray into matters of ultimate spiritual concern. But is such a mollifying attitude necessary in avowedly open societies? Can we not afford to take greater intellectual risks? Popper clearly thought so, which is why he treated instrumentalism as the counsel of those who lack guts and/or imagination.

Contemporary philosophy of science has not seen the problem of instrumentalism in quite this way, mainly because it has not seen it as a problem for instrumentalism – but for realism. This shift in the burden of proof is made possible once philosophers defined instrumentalism as explicitly neutral with regard to the theories that guide scientific research, resulting in a theory-neutral sense of 'empirical success'. The realist is then forced to defend the value added by science trying to do more than simply achieve and extend such empirical success. However, prior to Hume, it is difficult to find anyone other than Francis Bacon who thought that instrumentalism might provide an adequate account of science on such grounds alone (Laudan 1981, ch. 6). Even Bacon would have realized that the instrumentalist's notion of empirical success trades on a conflation of the mark of success and the means by which it was achieved – that is, a point about theory testing and a point about theory choice. After all, the record of empirical successes chalked up by a science remains the product of a particular history of inquiry (cf. Goodman's entrenchment theory), even after all traces of that process have been removed from official presentations of the science for purposes of further extending and applying its empirical base.

In this respect, as economists would say, instrumentalism is dedicated to masking science's 'path dependency', which in turn makes its accounts of science especially vulnerable to the grue paradox. Put another way, for Popper, instrumentalism encouraged a lack of historical reflexivity, what Hilary Putnam (1978) dubbed the 'pessimistic meta-induction' from the history of science. That is, if history is our guide, then the foundational explanatory theories of the sciences are likely to be superseded in a century's time, without necessarily undermining the cumulative character of their findings. On this basis, Popper's marching orders to scientists are clear: hasten history's course by speeding up the rate of criticism. However, these orders cannot be followed if instrumentalist philosophers discourage scientists from accentuating their theoretical roots, which open them to competing ways of organizing (much) the same data to point in different epistemological and ontological directions. Underlying Popper's sensibility here is an existentially rooted anti-foundationalism – namely, the belief that because we could have reached comparable levels of empirical success by alternative theoretical means, we should not fear losing our genuine epistemic achievements by radically changing our theoretical premises on the back of a falsified prediction or some other major empirical setback. Such an attitude permits an epistemic confidence that welcomes regular shifts in paradigms – in science *and* politics.

Anti-inductivism as one psychologist's love-hate relationship with psychology

In formulating his philosophical views, Popper may have been helped by not having been formally trained in physics. His Ph.D. was in educational psychology under the supervision of Karl Bühler, a pioneer in the experimental study of 'imageless thought', the subject matter of what we now call 'cognitive science', which is concerned with forms of consciousness oriented towards an object that is not present to the mind in the manner of a proxy observation (Berkson and Wettersten 1984, ch. 5). In the wake of Kripke (1977), we may think of such objects as making 'semantic' but not 'pragmatic' (or 'speaker's') reference. In other words, one seeks an object that satisfies a certain definite description or a solution that meets a specific set of disciplinary criteria without necessarily possessing a mental image from a previous encounter with the object of inquiry. Note that even if the object in question has not been previously encountered, nevertheless most if not all of its defining properties may have been. However, the properties emergent on their combination that uniquely identify the object remain unknown until the object is correctly encountered, which in turn marks the consummation of the act of thought. Popper's account of science as historically extended organized inquiry may be understood as a theory of imageless thought writ large. It provides a context for understanding his dogged refusal to assimilate 'verisimilitude' to subjectivist notions of probability, which are too strongly anchored in prior expectations (cf., e.g., Popper 1963, ch. 10).

For a precedent, when the French *philosophes* Turgot and Condorcet characterized the overall 'progressive' movement of history, the term they used was *tâtonnement*, based on how they understood how buyers and sellers agree prices so as to clear the market, a process that they believed was continually reiterated as the economy expanded (Rothschild 2001, ch. 6). A century later Léon Walras abstracted from their historical considerations to make *tâtonnement* the basis for general equilibrium theory, the cornerstone of neoclassical economics. A century after that, Pierre Teilhard de Chardin (1955) appropriated the same term for his version of creative evolution. *Tâtonnement* literally means 'groping in the dark'. It is easy to think of either a settled price (à la Turgot) or some 'omega' species (à la Teilhard) as the object in question. The idea also helps to explain Popper's use of the searchlight metaphor to capture the moment of experimental testability that reveals whether the sought object has been found (ter Hark 2009). It captured the 'world three' character of objective knowledge, which Popper explained as a long-term, largely unintended consequence of the exigencies facing our collective existence. We enter this realm of being, which transcends the worlds of both matter and mind, when we turn our attention to problems solutions to which are presupposed by our ability to solve the sort of real-world problems that involve the direct engagement of human designs with material outcomes (Popper 1972). In this way science was born – starting with mathematics as a sphere of inquiry independent of its applications (Fuller 1988, ch. 2).

Today's histories of experimental psychology tend to present the Würzburg school as a transitional stage between Wilhelm Wundt's original sensation-based version of introspective psychology, from which the Würzburgers revolted, and the more holistic but objectivist vision of psychology pursued by the Gestalt school, which most Würzburgers eventually joined. (For a good comprehensive philosophical history of the transition, see Kusch 1999, Part I.) As it turns out, Popper caught the Würzburgers at the peak of their influence, as Popper's doctoral supervisor, Karl Bühler, held the chair in experimental psychology at the University of Vienna just when Sigmund Freud's psychoanalysis, the discipline's private-sector competitor, had itself reached its peak of local influence (Ash 1987).

This bit of history helps to explain Popper's rather schizoid attitude to psychology as a discipline in the development of his anti-inductivist views. Never one to neglect youthful experience as a source of philosophical insight, Popper reported a conversation he had in 1919 (when aged 17) with Alfred Adler, then Freud's most publicly recognized disciple, whose ideas about the 'inferiority complex' and 'lifestyle' would dominate the reception of psychoanalysis in post–World War II America. At the time, just before going to university, Popper was working as a social worker for a Viennese inner-city youth project under Adler's auspices. What turned Popper off to psychoanalysis was the ease with which Adler could diagnose a child's problems by the simple matching of hearsay to past clinical experience without having examined child directly. Given that this conversation took place at a party in Adler's home, the rather earnest Popper may have overreacted to his host's bluff manner (Hacohen 2000, 92). Nevertheless, the episode became one of Popper's set pieces for explaining how he came to demarcate science from pseudoscience (cf., e.g., Popper 1963, ch. 1).

By all accounts, Adler was a purveyor of socially progressive views, including an egalitarianism in matters of class and gender that eventuated in his expulsion from Freud's circle. In this respect, his politics were quite close to that of the young Popper. However, Popper detected a clear difference between Adler's egalitarian ends and the inegalitarian means that he used to pursue them. In particular, Adler's exceptionally broad – and seemingly exceptionless – explanatory account of psychodynamic development amounted to a high-minded prejudice that circumscribed his interpretation of the child's response to treatment. Thus, a child who failed to respond well to improvements in his or her social environment would be diagnosed as engaged in resistance. There was no question of the original diagnosis having been at fault. The child previously stereotyped as irredeemable remained stereotyped, but now as redeemable under scripted conditions. Given the trust that both politicians and parents placed in Adler, his word carried enormous weight – all the more reason the young Popper found Adler's casual generalization of his clinical practice irresponsible.

Though little remarked, it is striking that when Popper wanted to stress the critical moment of scientific inquiry, the word that always came up was *risk*. Adler never placed his ideas at risk because he could not cede the epistemic privilege of his track record, which was after all the basis of his livelihood. A proper test of his track record to see whether it was based on science or superstition – that is, a genuine causal understanding of children's problems versus a series of lucky guesses – could result in an embarrassing outcome that would result in his losing reputation and hence clients. How much more convenient, then, it is to generate a pre-emptive feeling of success for any new case, so that if – or when – difficulties arise, social workers are ready to limit any damage to that expectation by excusing, marginalizing or papering over potentially discrepant outcomes.

Ethnomethodologists call this activity, in which the young Popper was loath to become involved, 'repair work' (Garfinkel 1967). Its prevalence in everyday life is often cited as evidence that the normative structure of society needs to be actively constructed on a moment-to-moment basis. On this basis, Wittgensteinians and ethnomethodologists have often made common cause in forging a philosophy-sociology alliance in science studies under the rubric of that protean term 'constructivism' (Lynch 1994; Sharrock and Read 2002). But whereas Wittgensteinians focus on the conceptual point that, at any moment, a social practice may be taken in any number of different directions, given that its track record may be justified by any number of theories, ethnomethodologists alight on the empirical observation that prior commitment to one such theory can be maintained in the face of any number of outcomes. In both cases, the 'normativity' of logic alone appears to pass for that of society, such that what is not logically prohibited is presumed to be socially permissible, so that one simply needs to see what happens in practice.

Such a position is normally dismissed as oblivious to any standards of conduct that the agents may have learned, let alone long-established power relations, that intercede between what is logically possible and socially permissible. When sociologists counterpose a 'structuralist' or 'macro' perspective to the individualist micro-orientation of ethnomethodology, this is what they mean. To be sure, some of Popper's animus can be explained this way, given his default sympathy for socialism, which led him to keep a studied distance from the more libertarian tendencies of the political economist Friedrich Hayek; yet the two of them, in their common opposition to totalitarianism, joined sides on a 'my enemy's enemy is my friend' basis (Hacohen 2000, 481–2). Not surprisingly, Popperians have excoriated the sort of sociology that would normalize, if not valorize, a process that systematically turns a blind eye to problems deeper than what social agents are normally willing to stomach to get on with each other (Gellner 1979, ch. 2). It is one thing to admit that the truth can never be determined with certainty, but quite another to encourage the studious avoidance of testing what one currently presumes to be true. The latter reduces politics to politeness, as tact passes for 'tacit knowledge'.

But Popper's objections to the probity of 'repair work' run deeper in ways that can be explained in terms of his early study of psychology. Popper believed that we are born holding many false ideas about the world, which only come to light as we conceptualize beyond what is necessary for our biological survival (Petersen 1984). In this respect, the epigraph to Popper (1963) is an instance of philosophical insight generated by ironizing irony – which is to say, taking the original statement straight. Popper starts *Conjectures and Refutations* with Oscar Wilde's 'Experience is the name that everyone gives to their mistakes'. While Wilde was clearly remarking on people's seemingly endless capacity for self-justifying repair work, Popper took the quip to mean that it is only by making mistakes that we acquire 'experience' in any epistemically meaningful sense. Everything else is simply operating by default, whereby responsiveness to the external world is, from an engineering standpoint, a redundancy designed to remind us of the original script that we are supposed to follow.

Put this way, Popper is making an ontological point, trying to draw a clear line in humanity's status as what, in the early twentieth century if not today, would be called *machine* and *organism*. A machine maintains itself by virtue of having been programmed to respond to various anticipated states of adversity, whereas an organism can alter the terms in which it confronts adversity, even if it cannot alter its own programme (cf. Rosen 1999, ch. 17). Thus, Joseph Agassi (1985), perhaps Popper's most faithful follower, has built an entire philosophy around the idea that a science turns into a technology once its horizons are limited to where it is reliably effective. Practitioners of such a risk-averse body of knowledge are acutely aware of their comfort zone (or conditions of applicability) and are determined to stay within them. The mystique of expertise in the public sphere is arguably constructed in just this way (Fuller 2002, ch. 3), which in turn implies that it can be experimentally deconstructed once experts are encouraged to engage in even slightly counterfactual speculations (Tetlock 2005). The machine-like character of expertise also helps to explain the instinctive Popperian revulsion to epistemic deference that has become so popular in analytic social epistemology (Diller 2008). The problem here is less to do with the 'experts' trying to maximize the applicability of their expertise than with the 'laity' falling for the bait-and-switch of thinking that their distinctive knowledge interests are adequately encompassed by any such expertise (Fuller 1988, ch. 12).

Metaphysically speaking, such mechanization becomes a problem once it starts to replace, rather than enhance, the organic dimension of our being. Before the Scientific Revolution, one might say, the problem was trying to get Christians to think of themselves as more than mere divine machines. The revolutionary breakthrough was to infer from a fairly literal biblical understanding of humans as creatures 'in the image and likeness of God' that we are also mechanics, just like God, capable of reverse-engineering reality back to first principles. The secular descendant of this attitude remains in the scientific impulse to reduce empirical complexity to the constrained application of a finite set of laws. In the aftermath of the Scientific Revolution, the main problem has been to ensure that we do not slip back from this state of species confidence, this time to an atheist version of the pre-revolutionary condition. It could still happen, in response to the massive risks we have incurred over the past four centuries in the course of reconfiguring our relationship to the environment. The signs are already there, as people claim there are built-in limits to our capacity for action that constitute what might be seen as our 'ecological competence'. To Popperian eyes, such a concept is worthy of severe test.

Conclusion: the problem of assessing Popper's fortunes

Popper pursued his philosophy of science with a seriousness that was admirable and/or obsessive. Regardless, it has survived primarily as a glorified rhetorical device – albeit one deployed to great effect, though not necessarily by Popper himself. Foundational experimental work on the presence of 'confirmation bias' in the psychology of both lay people and experts starting in the 1970s presumed falsifiability as the normative scientific attitude – which those experiments then proceeded to falsify (Tweney et al. 1981, Part IV). Around the same time, early sociologists of scientific knowledge made great sport of the fact that famous scientists bore witness to Popper's heroic image of them as 'permanent revolutionaries', while ethnographic studies of normal scientific practice revealed little evidence of falsification in action (Mulkay and Gilbert 1981). Notwithstanding these empirical embarrassments, some who remained convinced that the Popperian ideal captured the spirit of scientific inquiry started to question whether it could be realized in an individual human being. Perhaps a specifically organized social group or a computer programme might prove a more appropriate vehicle (Fuller 1993, Part III).

But even scientists who claim the efficacy of falsificationism in their own fields tend to infer – in a rather un-Popperian way – that because major past scientific errors and even misconduct could be reasonably traced to a failure to falsify one's pet hypotheses, it follows that falsificationism was behind those who escaped such ignominy and perhaps came to be celebrated. This false dichotomy is especially operative in popular presentations designed to demarcate 'science' from 'pseudoscience'. After all, it is usually not too difficult to show that purveyors of 'pseudoscience' have clever ways of shielding their hypotheses from direct refutation. But left unsaid is that practitioners of 'science' are usually no less adept. This pleasant superstition has been carried along by a rather loose sense of 'falsification', in which an accidental discovery might count as a 'falsification' even without specifying the prior theoretical claims that the discovery has supposedly contradicted. People are surprised all the time, sometimes significantly – but attributions of significance presuppose a context for making sense of the event. A surprising outcome that arises under surprising circumstances does not count. Thus, Popperian falsification requires explicit experimental stage setting, at the end of which one can state which one or more previously plausible hypotheses have now been excluded from further consideration.

Compounding the problem of assessing Popper's philosophy is that his own normative judgements about science, as well as those of his followers (especially Paul Feyerabend), tended to veer substantially from those of the scientific orthodoxy. In particular, they erred on the side of levelling the epistemic playing field. To be sure, a cautious falsificationist like Imre Lakatos (1970) worried that Popper's ethic made it too easy to eliminate new competing theories before they had a chance to develop. Nevertheless, as can be already seen in Popper's early encounter with

Adler, in practice Popperians have more often aimed their fire on established theories that tacitly incorporate dogmatic elements that render them unfalsifiable. Indeed, they have pursued pseudoscience in a spirit rather opposed to that of most other philosophers, who appear to start with agreed intuitions about what counts as pseudoscience and then differ over the principles of which they run afoul (Hansson 2008). In contrast, Popperians pursue their principles and let the cases fall where they may. In this way, even Lakatos (1981) himself ran roughshod on the scientific establishment's self-understanding by defining the task of the philosopher of science in terms of the 'rational reconstruction' of the history of science – the suggestion being that despite science's historic epistemic successes, it could have proceeded much more efficiently had it followed some sage philosophical advice.

In short, for all their pro-science attitudes, Popperians are inclined to bet against the scientific consensus. They are most definitely not Lockean 'underlabourers' (cf. Fuller 2000b, ch. 6). Indeed, Popper's former student from the London School of Economics in the 1950s, George Soros, the financier-turned-philanthropist, must count as the person who has most successfully internalized his anti-inductivist worldview. Trained in the arts of arbitrage, Soros has managed to stay in business by assuming that, for any commodity, half the market overvalues it and half undervalues it. This means that a profit can be made simply by falsifying both: that is, buying low and selling high. The question then is when the two opposing errors are sufficiently discrepant that one can make the biggest killing. The trick is to figure out what the philosophical version of Soros's winning strategy might look like for a revival of Popper's fortunes.

References

Adorno, T. (ed.) (1967), *The Positivist Dispute*. London: Heinemann.

Agassi, J. (1985), *Technology: Philosophical and Social Aspects*. Dordrecht: Kluwer.

Ash, M. (1987), 'Psychology and Politics in Interwar Vienna', in M. Ash and W. Woodward (eds), *Psychology in Twentieth Century Thought and Society*. Cambridge: Cambridge University Press, pp. 143–64.

Bateson, G. (1972), *Steps to an Ecology of Mind*. Chicago: University of Chicago Press.

Berkson, W. and J. Wettersten (1984), *Learning from Error*. La Salle, IL: Open Court Press.

Diller, A. (2008), 'A Popperian Theory of Testimony', *Philosophy of the Social Sciences*, 38: 419–56.

Fuller, S. (1988), *Social Epistemology*. Bloomington: Indiana University Press.

— (1993/1989), *Philosophy of Science and Its Discontents*, 2nd edn. New York: Guilford Press.

— (2000a), *The Governance of Science: Ideology and the Future of the Open Society*. Milton Keynes, UK: Open University Press.

— (2000b), *Thomas Kuhn: A Philosophical History for Our Times*. Chicago: University of Chicago Press.

— (2002), *Knowledge Management Foundations*. Woburn, MA: Butterworth-Heinemann.

— (2003), *Kuhn vs. Popper: The Struggle for the Soul of Science*. Cambridge, UK, and New York: Icon and Columbia University Press.

— (2011). *Humanity 2.0: The Past, Present and Future of What It Means to Be Human*. London: Palgrave Macmillan.

Galison, P. (1990), 'Aufbau/Bauhaus: Logical Positivism and Architectural Modernism', *Critical Inquiry*, 16: 709–52.

Garfinkel, H. (1967), *Studies in Ethnomethodology*. Englewood Cliffs, NJ: Prentice Hall.

Gellner, E. (1979), *Spectacles and Predicaments*. Cambridge: Cambridge University Press.

— (1998), *Language and Solitude*. Cambridge: Cambridge University Press.

Goodman, N. (1955), *Fact, Fiction and Forecast*. Cambridge, MA: Harvard University Press.

Hacohen, M. (2000), *Karl Popper, The Formative Years: 1902–1945*. Cambridge: Cambridge University Press.

Hansson, S. O. (2008), 'Science and Pseudo-science', *Stanford Encyclopedia of Philosophy*. http://plato.stanford.edu/entries/pseudo-science/.

Jarvie, I. (2001), *The Republic of Science: The Emergence of Popper's Social View of Science*. Amsterdam: Rodopi.

Kripke, S. (1977), 'Speaker's Reference and Semantic Reference', *Midwest Studies in Philosophy*, 2: 255–76.

Kuhn, T. S. (1970/1962), *The Structure of Scientific Revolutions*, 2nd edn. Chicago: University of Chicago Press.

Kusch, M. (1999), *Psychological Knowledge*. London: Routledge.

Lakatos, I. (1970), 'Falsification and the Methodology of Scientific Research Programmes', in I. Lakatos and A. Musgrave (eds), *Criticism and the Growth of Knowledge*. Cambridge: Cambridge University Press, pp. 91–195.

— (1981), 'History of Science and Its Rational Reconstructions', in I. Hacking (ed.), *Scientific Revolutions*. Oxford: Oxford University Press, pp. 107–27.

Laudan, L. (1981), *Science and Hypothesis*. Dordrecht: D. Reidel.

Lynch, M. (1994), *Scientific Practice and Ordinary Action: Ethnomethodology and Social Studies of Science*. Cambridge: Cambridge University Press.

Mulkay, M. and N. Gilbert (1981), 'Putting Philosophy to Work: Karl Popper's Influence on Scientific Practice', *Philosophy of the Social Sciences*, 11: 389–407.

Petersen, A. (1984), 'The Role of Problems and Problem Solving in Popper's Early Work on Psychology', *Philosophy of the Social Sciences*, 14: 239–50.

Popper, K. (1959/1935), *The Logic of Scientific Discovery*. New York: Harper and Row.

— (1963), *Conjectures and Refutations*. New York: Harper and Row.

— (1972), *Objective Knowledge*. Oxford: Oxford University Press.

— (1981), 'The Rationality of Scientific Revolutions', in I. Hacking (ed.), *Scientific Revolutions*. Oxford: Oxford University Press, pp. 80–106.

Putnam, H. (1978), *Meaning and the Moral Sciences*. London: Routledge and Kegan Paul.

Quine, W. V. O. and J. S. Ullian (1970), *The Web of Belief*. New York: Random House.

Rorty, R. (1979), *Philosophy and the Mirror of Nature*. Princeton, NJ: Princeton University Press.

Rosen, R. (1999), *Essays on Life Itself*. New York: Columbia University Press.

Rothschild, E. (2001), *Economic Sentiments*. Cambridge, MA: Harvard University Press.

Safranski, R. (2000), *Martin Heidegger: Between Good and Evil*. Cambridge, MA: Harvard University Press.

Sharrock, W. and R. Read (2002), *Kuhn: Philosopher of Scientific Revolution*. Cambridge: Polity.

Shearmur, J. and P. Norris (eds) (2008), *After the Open Society: Selected Social and Political Writings*. London: Routledge.

Stanford, P. K. (2006), *Exceeding Our Grasp*. Oxford: Oxford University Press.

Teilhard de Chardin, P. (1955), *The Phenomenon of Man*. New York: Harper and Row.

ter Hark, M. (2009), 'Popper's Theory of the Searchlight: A Historical Assessment of Its Significance', in Z. Parusniková and R. Cohen (eds), *Rethinking Popper*. Dordrecht: Springer, pp. 175–84.

Tetlock, P. (2005), *Expert Political Judgement*. Princeton, NJ: Princeton University Press.

Tweney, R., M. Doherty and C. Mynatt (eds) (1981), *On Scientific Thinking*. New York: Columbia University Press.

van Fraassen, B. (1980), *The Scientific Image*. Oxford: Oxford University Press.

CHAPTER 6

HISTORICAL APPROACHES:
KUHN, LAKATOS AND FEYERABEND

Martin Carrier

Historical approaches dominated philosophy of science as a whole between approximately 1965 and 1985. Thomas Kuhn's *Structure of Scientific Revolutions* appeared in 1962 and was first broadly discussed as a challenge to Karl Popper's falsificationist methodology at a conference held in London in 1965. The ensuing controversy between Kuhn and Popper and the later contributions from Imre Lakatos, Paul Feyerabend and others did much to fuel the debate and lifted theory change to the top of the agenda of philosophy of science. Two characteristics are essential for the historical approaches in question: first, the emphasis on the practice of science in its historical development (accompanied by a more detailed consideration of case studies from the history of science) and, second, the focus on theory evaluation and, in particular, theory comparison. Historical approaches chiefly aimed to uncover the reasons for theory change or the nature of scientific progress. Theory evolution and theory evaluation were the focal points of the historical approaches under consideration.

The background: cumulative history and falsificationism

The controversy about historical approaches unfolded against a backdrop that had shaped philosophical thoughts on scientific progress for quite some time. This cumulative view of progress suggests that theory change amounts to piling up truths (mostly about laws of nature). That is, what science has once recognized as true remains part of the system of knowledge forever. Science needs to retract theoretical claims to be sure, but such withdrawal is confined to illustrations of the laws or to intuitive models, for one, and to restrictions of the scope of laws, for another. The first item is highlighted by Moritz Schlick, among others. He claimed that no theory that has been verified will ever be jettisoned in its entirety. Rather, its key elements will be preserved by its successor theory. Only two sorts of changes are typically introduced by the successor: first, adding new details that serve to increase the match with the data and, second, removing misleading illustrations that had been introduced hastily so as to facilitate understanding and use of the theory (Schlick 1938, 203). Consequently, what has been established by scientific research is essentially left unaffected by subsequent scientific progress.

However, the domain of application of the laws of nature, as accepted by science, may turn out to be smaller than anticipated at their discovery. This second item was underscored by Pierre Duhem. Duhem argued that knowledge of the laws of physics is always tentative and incomplete and that, correspondingly, exceptions occur in experience. As a result, the pertinent law needs to be restricted to particular circumstances. Such restrictions in scope are increasingly recognized in the course of scientific progress (Duhem 1906, 141–4). For instance, special relativity and quantum theory have made us realize that the domain of Newtonian mechanics is bound to small velocities (as compared to the speed of light) and great lengths (as compared to the atomic scale), but the Newtonian theory is nevertheless valid within this reduced realm. Underlying the cumulative view is the assumption that the scientific method is suitable for uncovering true results. Scientific tests are taken to be so challenging that all accounts that pass them can be relied upon. The tough examination process and the demanding standards of acceptance are the reason why there are no substantial retractions in science.

At the surface, a much more agitated picture of scientific progress emerges from Popperian critical rationalism or falsificationism. Popper took seriously the deep ruptures that affected fundamental physics in the first three decades of the twentieth century. In his view, the acceptance of relativity and quantum theories involved abandoning the entrenched system of classical physics. Such episodes show that even venerable theoretical traditions may turn out to be ill-conceived. There is more to scientific progress than revealing unexpected limits of scope.

In contrast to the impoverished standard image of falsification-ism, Popper was well aware of the flexibility that theoretical systems contain for coping with recalcitrant data. He distinguishes between a logical and a pragmatic sense of falsification. Falsification in the logical sense refers to the contradiction within a set of theoretical claims and observation statements. Falsifiability in this logical sense is a property of a theoretical system as a whole; it says that if certain observation statements were adopted, the theoretical system would be untenable (Popper 1935, 52–3). However, Popper also acknowledges that 'conven-tionalist ploys', later called 'immunization strategies', could be invoked so as to avoid the refutation of particular assumptions. By appending unsupported auxiliary hypotheses or by adjusting the definitions of the pertinent concepts, specific theoretical claims could be rescued from the threat of contrary experience (Popper 1935, 15–6, 45–50).

Accordingly, Popper remains in broad agreement with a celebrated thesis of Duhem to the effect that the bearing of an anomaly on a particular part of a network of hypotheses cannot be assessed by rely-ing on logic and experience alone. Duhem's thesis says that different theoretical hypotheses can be held responsible for a given empirical problem, and different measures can be taken to fix the flaw. Additional considerations, subsumed by Duhem under the heading of '*bon sens*', guide scientists in picking a culprit.[1]

Popper acknowledges that in many cases theoretical principles can be neither proved nor disproved by experience. Seemingly anomalous data leave loopholes for scientists intent on saving such a principle from prima facie contrary evidence. However, as Popper argues, this strategy of rescuing theories from refutation fails to capture an important fea-ture of theory change. The advancement of science is characterized by the occasional abandoning of theoretical accounts. The latter are not defended come hell or high water. The phlogiston theory, the caloric theory and the ether theory, all well-entrenched in their time, were given up in response to empirical criticism. They could have been preserved but instead were dropped. In Popper's view, the underlying rationale is that continuing to adjust a troubled theory until it finally accords with recalcitrant data deprives us of the opportunity to learn from our mistakes. We can only realize that something is fundamentally wrong if we accept the verdict of experience and treat a theory as refuted although it is not, logically speaking. Falsifiability in this pragmatic sense is an attitude of scientists toward theories in the first place, not primarily a property of theories themselves (although resoluteness is of no avail if a theory fails to entail any specific consequences). Popperian falsifiability is hence chiefly produced by a methodological convention (Popper 1935, 22–3, 26, 49; 1963, 323–4; Lakatos 1970, 23).

Popper sees his methodology expressed in features of the growth of knowledge and patterns of theory change. Such patterns involve venturing bold conjectures and trying to refute them in serious and imaginative attempts. Science is at its best when expectations are thwarted. As a result, Popper rejects the Baconian methodological maxim that scientists should be free of prejudices in approaching nature. On the contrary, premature anticipations are essential for scientific progress, provided that such prejudices are severely tested. The key to scientific method is not waiving prejudices but controlling them by examining their empirical consequences (Popper 1935, 223–5).

Regarding theory change, Popper argues, on the one hand, that a theory replaces a refuted predecessor, which means that the two are incompatible; on the other hand, he also insists that the successor approximately preserves the earlier account if the circumstances are suitably restricted (Popper 1935, 199, 221–2). This reveals that Popper rejects the cumulative view in letter but retains it in spirit. It was only Kuhn and Feyerabend who abandoned this view in word and deed. Kuhnian scientific revolutions are characterized by a succession of theories that cannot be reconstructed as preservation in the limit (Kuhn 1962, 2–3, 92–8). Similarly, Feyerabend claims that scientific change does not proceed by embedding a theory in a more comprehensive one but by a complete replacement of the predecessor (Feyerabend 1962, 44–5). As a result, scientific progress does not merely extend the system of knowledge to new ground but also encompasses collapse and retreat.

Kuhn's paradigm theory

Kuhn opens his *Structure of Scientific Revolutions* with the commitment to link up philosophy of science with the history of science more intimately than before. If history is taken seriously it can 'produce a decisive transformation in the image of science by which we are now possessed' (Kuhn 1962, 1). Although Popper was not Kuhn's primary target of his reproach of historical superficiality, the ensuing Popper-Kuhn controversy served to make the relevant fault lines salient.

Kuhn introduced the distinction between a comprehensive theoretical framework and its more specific articulation. He labelled the framework a 'paradigm' and defined it to include theoretical principles, methodological or metaphysical commitments and a collection of exemplary solutions to problems (whence derives the designation 'paradigm'). For example, the paradigm of nineteenth-century wave optics proceeded from the assumption that light is to be conceived as transverse oscillations of a pervasive medium. Specific versions of the paradigm consisted in more concrete accounts of optical phenomena such as refraction, diffraction and interference. A scientific discipline that is dominated by one particular paradigm has entered the stage of 'normal science.' The shared commitment to an overarching framework makes it unnecessary for scientists to enter in-principle debates about the appropriateness of certain approaches and thus allows them to focus on more productive, technical work.[2]

In normal science, a paradigm has reached a monopoly and rules unquestioned. Its claims are not subjected to empirical examination. They are compared with experience, to be sure, but if the two don't match, the anomaly is not viewed as a shortcoming of the paradigm. The blame is rather attributed either to additional unrecognized influences or to a lack of creativity and skill on the part of the scientists. As a result, paradigms are immune to empirical counterinstances in normal science (Kuhn 1962, 77–80; 1970a, 6–7; Hoyningen-Huene 1989, 218–19).

Kuhn's picture of normal science stands in stark contrast with the Popperian maxim of testing theories severely. Popper's methodological advice to scientists is to take pains to actively produce counterinstances to a theory and to take them seriously in any event as potential refutations of the theory. By contrast, Kuhn's normal scientists are licensed to shelve unsolved problems and to go ahead undauntedly. The difference between Popper and Kuhn is not to be understood as the gap between lofty normative principles and a sloppy practice. Indeed, Kuhn gives epistemological reasons for the lenience toward anomalies he assumes to be characteristic of normal science. His claim is that every theory is fraught with anomalies, so that taking each such anomaly as a potential refutation would put an end to science (Kuhn 1962, 79–82). By contrast, the immunity paradigms enjoy in normal science provides a basis for the tenacious pursuit of theories which is in turn a necessary precondition for overcoming enduring challenges. In striking opposition to Popper, Kuhn considers it a characteristic of mature science that the corresponding scientific community abandons the critical discussion of the fundamentals of a discipline (Kuhn 1970a, 6). Mature science is distinguished by waiving Popperian severe tests.

However, when anomalies continue to pile up, confidence in the paradigmatic principles is weakened and finally lost. A 'crisis' emerges in whose course alternative theoretical options are considered and pursued. Such a crisis frequently gives rise to a 'paradigm shift' that is characteristic of a 'scientific revolution'. One of Kuhn's pivotal historical claims is that a theory is never dropped without a promising successor candidate. Abandoning a paradigm is tantamount to adopting a new one (Kuhn 1962, 77, 145–7). In contrast to the smooth development of normal science, Kuhnian revolutions amount to the wholesale replacement of a conceptual framework; they are non-cumulative in that they involve taking back achievements that were accepted as part of the system of knowledge before.[3]

The non-cumulative character of scientific revolutions becomes conspicuous in two features, namely, in changes of problems considered relevant and in the emergence of so-called Kuhn-losses.[4] The change of problems is unsurprising at first sight. After all, it conforms well to the traditional picture of scientific progress that problems are solved and new problems are thereby generated. Kuhn admits that problem changes of this kind appear in a revolution, but he stresses that an additional pattern of 'problem dissolution' emerges. This occurs when a problem is dismissed as ill-posed and misleading (Kuhn 1962, 103). For instance, one of the challenges of optical theory in the latter part of the nineteenth century was to determine the mechanical properties of the ether such that the known laws of light propagation would follow. After the acceptance of special relativity, the ether was dropped. Accordingly, the question as to the mechanical properties of the ether was not answered, but rather rejected as bereft of scientific significance.

Second, scientific revolutions frequently involve what is now called Kuhn-losses. A new paradigm may be accepted in spite of the fact that some of the former explanatory achievements are thereby lost. More specifically, some of the empirical problems solved before are reopened again. In this way, anomalies are produced by the paradigm shift itself. To be sure, Kuhn-losses are accepted by the scientific community only in small measure, but their mere existence vitiates any claim to the effect that the new paradigm reproduces all the explanatory achievements of its predecessor. One of Kuhn's favourite examples is taken from the Chemical Revolution. In the framework of the so-called phlogiston theory, a metal was regarded as a compound of a specific component (the 'calx') and phlogiston. Since phlogiston was assumed to be present in all metals, the theory could explain why they resembled one another to a much greater extent than the corresponding calces (oxides in modern terminology). Lavoisier's oxygen theory, by contrast, considered metals to be elementary and thus could not account for their similarity. The adoption of the oxygen theory made an empirical problem reappear that was considered settled before (Kuhn 1962, 132, 157, 170; 1970a, 20; 1977, 323).

As a result of these and other discrepancies, the old and the new paradigm are separated by a yawning chasm. Revolutions are non-cumulative; they involve a fundamental transformation in which the earlier account is not even approximately preserved by its successor. The contrast between theories separated by a revolution is far-reaching and unbridgeable (Kuhn 1962, 5–6, 97–110). At first glance, Kuhnian crises and revolutionary ruptures fit well with the Popperian image of science; after all, in these periods, theories are empirically tested and evaluated. In this vein, the Popperian view was advertised as the methodology for innovative science (Watkins 1970, 32, 36–7). Kuhn demurred and claimed rather that even scientific revolutions do not involve falsifications. No theory is rejected for the reason alone that it contradicts the evidence; instead, two or more theoretical options compete with one another, and the rejection of one is tantamount to the adoption of another one (Kuhn 1962, 145–7).

In sum, the most important historical claims entertained by Kuhn are that periods of normal science or paradigm monopoly are followed by revolutions or periods of theoretical pluralism. In the former periods, no significant theoretical innovation occurs; alternatives are only considered during the latter. A revolution involves the substitution, not the mere abandonment, of a paradigm. And the new paradigm cannot be reconstructed as containing the claims of its predecessor 'in the limit'; revolutions proceed in a thoroughly non-cumulative fashion.[5]

Lakatos's methodology of scientific research programmes

Lakatos attempted to save Popper's general approach by adjusting it in light of what Lakatos took to be Kuhn's insights into patterns of theory change, for one, and Kuhn's assaults on the rationality of science, for another. Lakatos's aim was to show that theory choice can be reconstructed as a process governed by methodological rules and that, correspondingly, theory change is much more rational and much less subject to merely subjective factors than Kuhn assumed.

Lakatos's basic unit of scientific progress is the 'research program', that is, a series of theories that share a 'hard core' and a 'positive heuristic'. The hard core contains the fundamental postulates of the program; it is retained throughout the program's period of active pursuit. The positive heuristic encompasses guidelines for articulating the program; it singles out significant problems and offers tools to their solution. One of Lakatos's historical assertions is that the development of an excellent program follows its positive heuristic and does not merely respond to conceptual and empirical difficulties (Lakatos 1970, 47–52, 68–9).

The hard-core postulates are irrefutable within the pertinent program; dropping them means abandoning the program. However, their irrefutability is not based on the firm ground on which they rest but rather on the commitment of the program's proponents to maintain them. Instead, anomalies are accommodated by adapting a 'protective belt' of auxiliary hypotheses. This protective belt is expanded, modified, restructured or replaced entirely according to the empirical needs set by the anomalies. Duhem's thesis (see sect. 1) entails that an empirical problem never bears directly on a specific theoretical hypothesis. Accordingly, there is always room left for the scientists to adhere to a hypothesis by redirecting the refuting force of an anomaly to assumptions from the protective belt, where required adjustments are made (Lakatos 1970, 48–50).

Modifications within the protective belt may concern, first, observation theories, second, initial and boundary conditions and, third, additional assumptions within the respective theory itself. Take the solar neutrino problem that plagued nuclear physicists and theoretical astronomers during the last decades of the twentieth century. The pertinent hard core is a theoretical model of the sun that contains accounts of the relevant fusion processes, the elements produced, the energy converted and so on. This model also predicts that a certain amount of neutrinos is emitted by the sun. The flux of solar neutrinos that reach the earth can be measured, and the result was a third of the expected value. One could argue, first, that the procedures for detecting neutrinos are less efficient than assumed. This line entails that the right amount of neutrinos is emitted and that the fault lies with the detection process or the pertinent observation theories. Second, modifications at the level of initial and boundary conditions could be made by assuming a lower temperature in the interior parts of the sun. In such a case, a reduced amount of neutrinos would be emitted. A third option is to assume that the different types of neutrinos can be converted into one another. The measuring procedures were only able to detect electron neutrinos, that is, one of the three relevant types, so that a sort of detection leakage would be responsible for the missing neutrinos. The issue was resolved in 2001 in favour of the third option of neutrino oscillations. Until that point scientists had ample room for shifting the blame for the anomaly across the protective belt.

The hard core and the positive heuristic constitute the program's identity. A specific version of a program encompasses these invariant elements together with the changing protective belt. A research program unfolds as a chain of subsequent versions, each of which arises from its predecessor by some theoretical modification, usually performed within the protective belt. Lakatos appeals to methodological standards to distinguish between acceptable and inappropriate theoretical changes. These standards primarily stipulate which theoretical changes within a program are methodologically sound.[6] An acceptable successor version within a program is required, first, to remain in agreement with the positive heuristic of the program, second, to account for all those phenomena that are successfully explained by its predecessor – albeit possibly in a different fashion – and, third, to successfully predict some novel, hitherto unexpected facts. Lakatos demands the reproduction of the previous empirical achievements and the anticipation of observational regularities that were unknown to science before. The successful prediction of novel facts is the chief display of scientific quality. Rival research programs are to be evaluated analogously. Programs that satisfy these program-internal standards are compared by applying the conditions of reproduction and anticipation to the competitors. A superior program has to reproduce the explanatory successes of the rival program and predict effects unexpected within the framework of the latter. The empirical confirmation of theoretical anticipations is the stuff of which methodological superiority and scientific progress is made (Lakatos 1970, 33–6, 47–52, 68–9; see also Popper 1963, 326–9, 334–5; Carrier 2002, 59–63; 2008a, 279).

The demand to reproduce the explanatory achievements of another theory (a preceding version or an alternative program) does not amount to the requirement to reproduce the explanations themselves. Retention of the principles and the modes of derivation is not necessary; the relevant phenomena may well be accounted for in a disparate fashion. For this reason Lakatos's condition of progress is compatible with non-cumulative change in the Kuhnian sense of a complete replacement, rather than approximate retention, of a theory by another one.

Lakatos is chiefly concerned with the methodological explanation of theory choice. Kuhn had argued that during revolutionary periods the competitors advance different standards for judging the appropriateness of problem solutions. As a result, each competitor appreciates its own assets and its rivals' liabilities drawing on its own specific measures of adequacy. Naturally enough, the partisans of contrasting approaches fail to convince one another.[7] One of Kuhn's examples is again taken from the Chemical Revolution. Within the phlogistic framework, it was considered the chief task of chemical theory to account for the properties of chemical substances (such as hardness, combustibility, volatility and the like) along with their changes during chemical reactions. Consequently, chemical explanations are to be judged according to their ability to provide such an account. As a result of the switch to the oxygen theory, these problems moved into the background, whereas the challenge to accommodate reaction weights was considered pivotal. Consequently, the standards for judging the achievements of chemical theories were altered as well (Kuhn 1962, 107; 1977, 335–6).[8]

Lakatos shifts Kuhn's theory-specific standards into the positive heuristic. Since the heuristic determines which kinds of theoretical means and procedures are acceptable within a program or which theoretical aims are to be followed, theory-specific standards have a legitimate place there. In any event, the dependence of such standards on and their variation with particular theories square well with the heuristic nature of the standards. However, in contrast to Kuhn, no theory-specific standard is granted any influence on the comparative evaluation of programs (Lakatos 1970, 68). After all, compliance with the positive heuristic is only required within a program, and the judgement of entire programs relies exclusively on overarching, or 'transparadigmatic,' criteria.

Lakatos's methodological criteria specify requirements for the facts a theory can rightly count in its favour. His criteria operate by singling out facts explained in a particularly demanding manner, namely, by successfully predicting a novel fact at the same time. The appraisal of a theory is based exclusively on these successful predictions, not on all of its correct empirical consequences. If consideration is restricted to such a selected group of outstanding explanations, it is much easier to reach an agreement on how well the theory is doing empirically (Lakatos 1970, 36; 1971, 112).

Lakatos's methodology was formulated as a response to Kuhn's, and it leaps to the eye that some of Lakatos's concepts are modelled on Kuhn's notions (Kuhn 1970b, 256). A research program roughly corresponds to a paradigm, and a program change resembles a scientific revolution. The retention of the hard core and the positive heuristic in pursuing a program reproduces the continuity of normal science. Finally, the methodological license to give up accepted explanations and to accommodate the relevant phenomena in a different way captures the non-cumulative character of revolutions.

However, Lakatos's methodology does not simply reproduce Kuhn's notions in a different conceptual framework. It aims rather at meeting methodological challenges raised by the paradigm theory. A pertinent issue is Kuhn's claim about the immunity of paradigms to anomalies. This immunity looks like an irrational trait from a Popperian angle. Its analogue for research programs can be derived as follows. Supporting facts are constituted by correctly predicted empirical regularities. Such facts have not been accounted for by the competing program; they would otherwise not be novel. Thus, Lakatos's condition of progress suggests that only those facts which cannot be explained by its rival are suitable for buttressing a program. Conversely, only those facts militate against a program that favours its competitor. It follows that only those anomalies which can be solved by the rival in a qualified fashion (i.e., by predicting unknown phenomena at the same time) will count as failures (Lakatos 1970, 37). This entails that the mere inability to accommodate an observation does not bring a program into trouble (Lakatos 1970, 92). This conclusion agrees with the Kuhnian immunity claim so that the latter follows from Lakatos's requirements for successful explanations. Research programs are rightly immune to mere anomalies (Carrier 2002, 63–5).

Another historical pattern stressed by Kuhn is that paradigms are never dropped unless a potential successor is waiting in the wings. Scientific revolutions proceed by theory substitutions. Its Lakatosian analogue is No program abandonment without program replacement. A program is only discredited methodologically if a superior competitor is available. This condition can be derived from a corollary to the aforementioned immunity argument. This argument says that the liabilities of one theory are the assets of the other. There are no significant failures without an alternative solution. And obviously enough, if a theory is not in trouble, it should not be given up. It follows that a program can never be rated as deficient unless there is a contender attacking it with some success. Disqualification of a program is brought about by a corroborated rival (Lakatos 1970, 35; 1971, 112–13).

Kuhn levels his claims of paradigm immunity and paradigm substitution as descriptive objections to Popper's methodological requirements. Within the framework of Lakatos's methodology, by contrast, the two features of immunity and substitution constitute theorems rather than objections. They follow from Lakatos's conception of how theories are to be evaluated. Accordingly, if theories develop in the Kuhnian way, no conflict with methodological rules arises. Instead, Lakatos's conception provides a methodological explanation of these patterns of theory change (Carrier 2002, 65).

However, Lakatos's portrayal of scientific change does not simply reproduce Kuhn's picture but deviates from the latter in some respects. For instance, Lakatos's methodology entails that science exhibits a thorough theoretical pluralism rather than a monopolistic rule of paradigms punctuated by occasional cataclysms. Further, Lakatos's methodology rules out Kuhn losses since Lakatos demands that theories are not superseded until all the phenomena they had explained are also accounted for by the successor (see above). Regarding pluralism, the verdict of the history of science points rather in favour of Lakatos, while regarding Kuhn losses, it rather agrees with Kuhn.

Feyerabend's methodological anarchism

Paul Feyerabend's position was formulated mostly in response to Lakatos, who Feyerabend took to be the most sophisticated rationalist in matters of methodology and whose position Feyerabend intended to contrast with his relativistic stance.[9] Regarding science, Feyerabend's 'methodological anarchism' entails that there is no way to discern epistemic achievements of different theories. No account is justifiably superior to any other. The only defensible principle regarding theory assessment is 'anything goes' (Feyerabend 1975a, 27–8). Theories are to be welcomed as long as a plurality of them compete with each other and none gains a dominant position (Feyerabend 1975b, 365).

Pluralism is threatened if cognitive positions are given up so that the spectrum of approaches is narrowed. Consequently, Feyerabend's objective is to show that rejecting such approaches is never justified. I focus on two of Feyerabend's arguments regarding science. It is unwarranted to exclude a theory from further consideration on the ground of, first, external incoherence or its contradiction with other accepted theories and, second, discrepancies between theory and evidence. External coherence demands that different theories should be in agreement with one another and thus give rise to a unified picture of a realm of facts. However, as Feyerabend claims, there is no good reason to drop a theory because it contradicts other, entrenched accounts. The argument is epistemological and says that error is discovered by comparison, that is, by using a contrasting approach, not by analysis, that is, by confining scrutiny to a given approach. Weak spots are more easily revealed by bringing to bear a discrepant framework, whereas internal criticism always accepts some of the premises at hand. As a result, seemingly inferior ideas should be retained as a challenge for the dominant view. Deficient accounts should not be given up but rather be strengthened.

A theory grows and matures by mastering such challenges (Feyerabend 1975a, 29–32; 1975b, 361–2).

All advocates of historical approaches support pluralism (although, perhaps, restricted to particular stages of theory development), but most regard it as a heuristic tool for promoting the growth of knowledge. Pluralism is seen as a temporary feature that is apt to produce a more sophisticated consensus than would have been accessible by pursuing just one approach. Distributing epistemic resources across a variety of pathways is the best strategy for arriving at the optimum solution. Feyerabend dissents and requires pluralism to be permanent. The dominance of any theory is pernicious to science – even if the theory should be true. If the truth need not face contenders, the reasons for its superiority increasingly fall into oblivion. The truth is of no avail if it is mindlessly reproduced and eventually no longer understood. Pluralism is not a means for propelling the growth of knowledge; it rather represents the ideal state of knowledge. As a result, external incoherence is a methodological asset rather than a liability (Feyerabend 1975a, 47, 52; 1975b, 365).

A pivotal standard for judging theories is empirical adequacy. If the facts militate against a theory, it deserves to be given up. However, as Feyerabend puts forward, the facts may be dependent on theory. What he has in mind here is the theory-ladenness of observations: measuring procedures serve to link up indicator phenomena with theoretical states, and these links are established and accounted for by theory. However, if such observation theories are mistaken, a systematic observational error may arise (Carrier 1994, 10–19; 2008b, 69–77). Anomalies may be produced by erroneous observation theories. Again, as Feyerabend emphasizes, such mistakes are best disclosed by approaching the matter from a variety of viewpoints, in this case by employing different observation theories.

Feyerabend's favourite example concerns Galileo and the motion of the earth. Within the framework of the then-accepted impetus physics, the motion of the earth should have had observable effects on the terrestrial motion of bodies. The absence of such effects seemingly ruled out any diurnal and annual motion of the earth. However, Galileo adopted a different stance. He recognized that the apparent counterevidence was produced by appeal to the impetus principle: that all motion comes to rest unless maintained by a force. The phenomena were interpreted by using this principle as observation theory. Yet if the principle of inertia was invoked for this purpose, the anomalous evidence disappeared. This example shows that recalcitrant data are no good reason for dropping a theory. Rather, the trouble may be due to a mistaken observation theory (Feyerabend 1970, 204–5; 1975a, 64–7, 74, 98).

Feyerabend subsumes the lesson to be drawn from these considerations under two principles. The *principle of tenacity* licenses continued adherence to a theory in spite of known difficulties and anomalies. The *principle of proliferation* recommends addressing such problems by elaborating alternative theories. Feyerabend acknowledges that both principles are part of Kuhn's paradigm theory. However, Kuhn restricts the appropriateness of each principle to particular stages of the development of science. Tenacity matches the complacency of normal science; proliferation is the order of the day when winds of change are blowing. In contradistinction, Feyerabend would have it that the two principles represent different aspects of scientific method and are employed jointly (Feyerabend 1970, 203–8, 211).

Feyerabend's criticism is mostly directed at a ruthless falsificationism that is, moreover, blind to the room left for fixing failures of a theory (that is expounded by Duhem's thesis). No proponent of the theory-change tradition, and no philosopher of science in the twentieth century at that, has ever advocated such a position. The conclusion is, accordingly, that Feyerabend overstates his case. It is true, theories can never be judged unambiguously, but they may pass or fail demanding empirical tests and thus exhibit strongly unequal degrees of epistemic virtues like explanatory and predictive power. Feyerabend's considerations fail to convincingly support the relativistic stance he adopts.

Essentials of historical approaches

Although the differences in detail among the various advocates of historical approaches to methodology are conspicuous, they also exhibit a large common ground. First, they share the appreciation of historical detail. The idea is that placing science in its actual complexity at the focus of analysis is apt to give rise to a richer and more adequate account of methodology than the preceding practice of alluding superficially to schematic examples. Second, methodology is supposed to elucidate the process of comparative evaluation of theories and the subsequent selection of one account at the expense of alternatives. The way of choosing between theories in a certain problem-situation is the key to methodology and to the rationality of science as a whole (Popper 1963, 335). The adoption of this framework leads to an emphasis on theory change and to the acceptance of pluralism within science. This approach goes along with abandoning the cumulative view of history.

A prima facie distinction among the leading advocates concerns the bearing of methodological rules and criteria. On the face of it, Popper and Lakatos opt for the rule of law regarding the assessment of the epistemic achievements of theories, whereas Kuhn and Feyerabend want to leave this assessment to the discretion of individual scientists.[10] However, the boundaries between the two camps become fuzzy on closer scrutiny. Popper emphasizes that judgement is necessary for resolving issues of refutation (Popper 1935, 16; 1963, 64–5), and Lakatos stresses that his tough standards are more appropriate for rationally reconstructing theory change than for directing it (Lakatos 1971, 117, 131–4). When it comes to the nitty-gritty, all parties to the dispute agree that the import of general methodological rules on the development of science is limited.

Another point of contact among the proponents of the theory-change tradition is the adoption of a pronounced normative stance. Even Kuhn, who insisted on the descriptive adequacy of methodological accounts, also claimed that the pattern of theory change he suggested is epistemically appropriate. That is, research pursued along Kuhn's lines stands a good chance of contributing to the growth of knowledge (Kuhn 1962, 205–6; 1970b, 237). Popper's normative approach is made most conspicuous by his condemnation of Kuhn's normal scientist as a deplorable figure. Normal scientists exist, but for the betterment of science they ought not (Popper 1970, 53). Lakatos distances himself from what he considers Kuhn's merely psychological account of theory change and insists on keeping epistemological elements in the methodology (Lakatos 1970, 90–1). Finally, Feyerabend opposes Kuhn's account because he takes normal science to be a dull and complacent enterprise that even exhibits anti-humanitarian tendencies (Feyerabend 1970, 210). The bottom line is that for all positions under review here, theory change is characterized with reference to the epistemic aims of science; science contributes first and foremost to the growth of knowledge. It is not alone the driving forces of theory change that are intended to be illuminated, but the good reasons that govern the process.

Finally, the mainstream of the theory-change tradition attempted to steer a middle course between realism and relativism.[11] Epistemic progress in science becomes manifest in objective achievements such as broad scope of theories and the increased precision of explanations based on them. But there are no convincing reasons for believing that science eventually will manage to represent the blueprint of the universe (Kuhn 1962, 205–6; Lakatos 1970, 99–101; Laudan 1984). Underlying this instrumentalist restraint is the commitment to the underdetermination of theories by evidence and the emphasis on the role of imagination in science. The facts always leave room for different theoretical accounts, and the pathways of science are paved by human creativity rather than dictates of nature. Science is and remains a human creation. The more recent debates about methodological matters tend to narrow the issue to the dichotomy of realism and relativism. It might help to recognize that the historical approaches had elaborated a position in between, namely, the notion of objective but non-representational epistemic merit.

The hallmark of the historical approaches is the connection established among theory choice, scientific progress and the rationality of science. Theory choice involves the adoption of a cognitively superior account whose credentials are specified or made explicit by methodological approaches. These criteria refer to the relations between the theoretical claims entertained and the observations made. The sequence of the theories adopted by the scientific community constitutes scientific progress, and the methodological criteria suggested are judged by their ability to capture essential elements of this development. The ability to reconstruct theory change as evolution in accordance with methodological judgement buttresses the claim that theory change proceeds rationally. Certain patterns of change constitute scientific rationality and scientific progress.

Historical approaches set a counterpoint to abstract reasoning in matters of confirmation that cranks concrete evaluations of theories out of overarching goals of science, such as high probability, and becomes manifest, in particular, in all sorts of inductive logic (such as Carnap's confirmation theory or latter-day Bayesianism). Historical approaches attempt to include to a much greater extent the variegated details of the actual practice of science as it unfolds in history. Historical approaches agree with social constructivism in taking social factors seriously, but unlike the latter, they emphasize that science advances in epistemic respects. The primary aim is the elucidation of the process of the growth of knowledge in the sense of propositions of superior cognitive quality and empirical reliability.

Notes

1. Duhem 1906, 152–4, 175–6. However, Popper, disagreeing with Duhem, claims that 'in quite a few cases' the mistaken hypothesis can be identified as the one that was necessary for deriving the refuted prediction (Popper 1963, 324).
2. See Hoyningen-Huene 1989, 133–62 for Kuhn's notion of 'paradigm'.
3. Bird 2000, 30–48, gives a succint overview of Kuhn's notions of normal and revolutionary science.
4. See Carrier 2002, 55–57 for additional relevant features.
5. For a critical discussions of Kuhn's historical claims see Bird 2000, 49–63.
6. Lakatos oscillated between different versions of his criteria. I confine myself to what I take to be the original version. For alternative readings see Carrier 2002, 69.
7. Kuhn 1962, 109–110. Such differences in the standards of evaluation have come to be called 'methodological incommensurability,' in contradistinction to 'semantic incommensurability,' that is, non-translatability due to deep theoretical contrasts involved. Neither Kuhn nor Feyerabend initially distinguished between the two sorts of incommensurability (see Carrier 2001).
8. Kuhn later emphasized the importance of comprehensive standards that transcend single paradigms, but claimed that they are unable to guide methodological judgement unambiguously. This 'Kuhn-underdetermination' plays no role in the debate under consideration (Carrier 2008a, 274–8).
9. In fact, Feyerabend attempts to show that Lakatos's criteria are vacuous and that Lakatos is as relativistic implicitly as Feyerabend is explicitly (Feyerabend 1975b, 215–19). Feyerabend's *Against Method* is dedicated to 'Imre Lakatos. Friend and fellow anarchist.'
10. For Kuhn, the pertinent leeway exists only in revolutionary periods, but outside of such periods no theory choice is made anyway.
11. The two exceptions are Popper, the realist, and Feyerabend, the relativist.

References

Bird, Alexander (2000), *Thomas Kuhn*. Princeton, NJ: Princeton University Press.

Carrier, Martin (1994), *The Completeness of Scientific Theories. On the Derivation of Empirical Indicators Within a Theoretical Framework: The Case of Physical Geometry*. Dordrecht: Kluwer.

— (2001), 'Changing Laws and Shifting Concepts: On the Nature and Impact of Incommensurability', in P. Hoyningen-Huene and H. Sankey (eds), *Incommensurability and Related Matters*. Dordrecht: Kluwer, pp. 65–90.

— (2002), 'Explaining Scientific Progress. Lakatos's Methodological Account of Kuhnian Patterns of Theory Change', in G. Kampis, L. Kvasz and M. Stöltzner (eds), *Appraising Lakatos: Mathematics, Methodology, and the Man*. Dordrecht: Kluwer, pp. 53–71.

— (2008a), 'The Aim and Structure of Methodological Theory', in L. Soler, H. Sankey and P. Hoyningen-Huene (eds), *Rethinking Scientific Change and Theory Comparison: Stabilities, Ruptures, Incommensurabilities?* Dordrecht: Springer, pp. 273–90.

— (2008b), *Wissenschaftstheorie: Zur Einführung*. Hamburg: Junius.

Duhem, Pierre (1906/1981), *La théorie physique – son objet, sa structure*. Paris: Librairie philosophique J. Vrin.

Feyerabend, Paul K. (1962/1981), 'Explanation, Reduction and Empiricism', in *Realism, Rationalism and Scientific Method (Philosophical Papers 1)*. Cambridge: Cambridge University Press, pp. 44–96.

— (1970), 'Consolations for the Specialist', in Imre Lakatos and Alan Musgrave (eds), *Criticism and the Growth of Knowledge*. Cambridge: Cambridge University Press, 197–230.

— (1975a/1987), *Against Method. Outline of an Anarchistic Theory of Knowledge*. London: Verso.

— (1975b/2006), 'How to Defend Society Against Science', in E. Selinger and R. P. Crease (eds), *The Philosophy of Expertise*. New York: Columbia University Press, pp. 358–69.

Hoyningen-Huene, Paul (1989), *Die Wisenschaftsphilosophie Thomas S. Kuhns*. Brunswick: Vieweg.

Kuhn, Thomas S. (1962/1996), *The Structure of Scientific Revolutions*. Chicago: University of Chicago Press.

— (1970a), 'Logic of Discovery or Psychology of Research?', in Imre Lakatos and Musgrave (eds), *Criticism and the Growth of Knowledge*. Cambridge: Cambridge University Press, pp. pp. 1–23.

— (1970b), 'Reflections on My Critics', in Imre Lakatos and Alan Musgrave (eds), *Criticism and the Growth of Knowledge*. Cambridge: Cambridge University Press, pp. 231–78.

— (1977), 'Objectivity, Value Judgment, and Theory Choice', in *The Essential Tension. Selected Studies in Scientific Tradition and Change*. Chicago: University of Chicago Press, pp. 320–39.

Lakatos, Imre (1970/1978), 'Falsification and the Methodology of Scientific Research Programmes', in John Worrall and G. Currie (eds), *Imre Lakatos. The Methodology of Scientific Research Programmes (Philosophical Papers I)*. Cambridge: Cambridge University Press, pp. 8–101.

— (1971/1978), 'History of Science and Its Rational Reconstruction,' in John Worrall and G. Currie (eds), *Imre Lakatos. The Methodology of Scientific Research Programmes (Philosophical Papers I)*. Cambridge: Cambridge University Press, pp. 102–38.

Lakatos, Imre and Alan Musgrave (eds) (1970/1974), *Criticism and the Growth of Knowledge*, Cambridge: Cambridge University Press.

Laudan, Larry (1984), 'Explaining the Succes of Science: Beyond Epistemic Realism and Relativism', in J. T. Cushing, C. F. Delaney and G. M. Gutting (eds), *Science and Reality. Recent Work in the Philosophy of Science. Essays in Honor of Ernan McMullin*. South Bend, IN.: University of Notre Dame Press, pp. 83–105.

Popper, Karl R. (1935), *Logik der Forschung*, Tübingen: Mohr, 1976.

— (1963), *Conjectures and Refutations. The Growth of Scientific Knowledge*. London: Routledge, 2005.

— (1970), 'Normal Science and Its Dangers', in Imre Lakatos and Alan Musgrave (eds), *Criticism and the Growth of Knowledge*. Cambridge: Cambridge University Press, pp. 51–58.

Schlick, Moritz (1938), 'Form und Inhalt. Eine Einführung in philosophisches Denken', in *Philosophische Logik*, ed. B. Philippi, Frankfurt: Suhrkamp, 1986, pp. 110–222.

Watkins, John (1970), 'Against Normal Science', in Imre Lakatos and Alan Musgrave (eds), *Criticism and the Growth of Knowledge*, Cambridge: Cambridge Univesity Press, pp. 25–37.

Worrall, John and G. Currie (eds) (1978), *Imre Lakatos. The Methodology of Scientific Research Programmes (Philosophical Papers I)*. Cambridge: Cambridge University Press.

CHAPTER 7

THE CONTINGENCY OF THE CAUSAL NEXUS: GHAZALI AND MODERN SCIENCE

Arun Bala

Introduction

Modern empiricism assumes that the first principles of any science can only be justified by appeal to experience and not by reason alone. Although it is widely acknowledged that this epistemological reorientation is one of the main factors that took modern science beyond ancient Greek science, it is not often recognized that the shift is the outcome of a long history of debate and controversy within Arabic philosophy between the school of *falsafah*, associated with Greek philosophical rationalism, and the school of *kalam*, associated with Islamic theology. Especially overlooked is the significant role played by the Arabic philosopher Abū Hāmed Muhammad ibn Muhammad al-Ghazālī (1058–1111), known as Algazel in the medieval West, in bringing about this epistemological revolution. Ghazali's contribution to the philosophy of modern science has been obscured not only by those who condemn him for initiating the decline of science in the Islamic world by his attacks on Greek rationalism represented by the *falsafah*, but also those who praise him for saving religion by exposing the limits of scientific rationalism. But neither those who vilify him for subverting science nor those who valorize him for saving religion seem to appreciate the role his critique of Greek science played in paving the way for modern science. We will find that although he did not propose the empiricist epistemology which underpins modern science and separates it from the Aristotelian and Platonic tradition that preceded it, the new epistemology could not have emerged without his attack on the rationalist foundations of Greek science.

Modern science sees the first principles of a scientific discipline as postulates or conjectures that are justified by their capacity to generate predications that can be confirmed by observation and experiment. By contrast, ancient Greek tradition – both Platonic and Aristotelian – assumed that the first principles had to be justified by appeal to their self-evidence. These principles had to be such that they could be apprehended by the intellect as indubitable. The main difference between Platonists and Aristotelians lay, not in how they were justified, but in how they were discovered. For Platonists they could only be discovered through the rational method of dialectics. By contrast Aristotelians stressed the role of prior experience in coming to know these first principles, although once apprehended their truth would be self-evident by virtue of being indubitable.

The difference in the Platonic and Aristotelian approaches to discovering first principles is rooted in their different metaphysical conceptions of what the being of a thing is. Plato maintained that the objects in the world are embodiments, albeit imperfect, of ideal forms in a transcendental realm. Hence these forms could only be intuited by the intellect, not perceived through the senses. Aristotle saw the forms as only existing in particular things and having no independent existence. Hence they could only be known through empirical observation of particulars, so that perceptual contact with objects is needed to give knowledge of the first principles of a science. However, Aristotle agreed with Plato that these first principles, once apprehended, would appear self-evident – that is, they would be indubitable without any possibility of being false.

It was the rationalist foundations of Greek science – especially associated with the Aristotelian tradition – that Ghazali set out to deconstruct. He largely succeeded, and as we will find, his success paved the way for the empiricist reorientation of modern science. But the significance of his contribution is often not appreciated, and to a large extent the blame for this lies in the prevalent Eurocentric history of modern science that has served as the context for understanding its philosophical basis. Eurocentric history assumes that the significant philosophical and theoretical influences on modern science are directly rooted in ancient Greek science. Although there is much historical evidence which has accumulated over the last five decades to render questionable this perspective, it nevertheless continues to shape our thinking. It suggests that non-European philosophical traditions cannot be relevant for current philosophy of science or for understanding its historical origins. This point has been made by Sundar Sarukkai:

> One reason to explain this complete denial of other philosophical traditions in mainstream philosophy of science is based on the belief that modern science is a product of Western civilization and hence any analysis of it is best done by the Western philosophical traditions of the same civilization. This is the originary issue for the philosophy of science. This particular point of view is further strengthened by the common observation that natural science grew out of philosophy. However, this observation emphasizes common origins of particular kinds of intellectual activities while ignoring the reasons as to why science broke away from traditional philosophy. In the history of science, there have been more attempts to show that the distinctive nature of scientific character was radically different from philosophy than attempts to show they were similar. This is a potential paradox for philosophy of science: it insists on using philosophical insights from traditions whose rejection in the first place catalyzed modern science.[1]

Sarukkai's observation is particularly relevant here. If modern science developed out of the philosophical critique of ancient Greek science, which it displaced, it is indeed paradoxical that philosophers of science today should treat ancient Greek philosophy, which constituted the foundation for Greek science, as a suitable basis for illuminating modern science. Moreover, if *kalam* developed as a counterpoint to the *falsafah* tradition in Arabic thought, and *falsafah* is deeply rooted in Greek philosophy, there are even stronger grounds for wondering if *kalam* played any role in the epistemological shift from Greek to modern science.

There are those who argue that *kalam* is not a tradition of philosophy at all, since it is motivated by theological concerns. However, appeal to such grounds to deny *kalam* the status of philosophy would even impugn major thinkers associated with *falsafah*, such as Ibn Sina and al-Farabi, since they used philosophical argumentation to defend their religious faith. Hence, if we are prepared to concede that *falsafah* is an attempt to naturalize Greek philosophical views within the matrix of Quranic revelation, we should also be prepared to concede that *kalam* is an effort to naturalize an alternative philosophical orientation into the same spiritual world. Indeed, if we repudiate giving philosophical status to *kalam* on the grounds of its theological interests, we would not only have to reject claims that Augustine and Aquinas in Europe made contributions to philosophy, but also repudiate much that is now taken to be a part of Hindu or Buddhist philosophy. Since we are prepared to extend the term *philosophy* to theological concerns in cultures beyond the Islamic world, provided they use argument and reason to justify their claims rather than mere appeal to scriptural authority, we should be prepared to extend the same courtesy to *kalam*.[2]

Indeed, refusing to treat *kalam* as an alternative philosophical tradition within Arabic civilization engaged in an extended competition with the *falsafah* would handicap our attempts to even understand the key ontological and epistemological issues that confronted the latter. Many of these concerned differences that separated the two traditions in their attempts to reconcile their philosophical reflections with Quranic revelation. We would also be squeezing their approaches to fit into quite different conceptions of philosophy associated with modernity – conceptions alien not just to *kalam* but also the *falsafah*. It may even hinder our efforts to understand the key concerns that separated the *falsafah* and *kalam* and conditioned their development and growth in dialogical conflict and engagement with each other.

Ghazali's critique of epistemological rationalism

Let us now examine more closely Ghazali's objections to the Greek inspired *falsafah* tradition in the Arabic world. The *kalam* tradition of medieval Islamic-Arabic thought he espoused developed as an antithetical tradition to the *falsafah* in Arabic civilization. Distinctive to *kalam* are its ontology of atomism and epistemology of causality developed in opposition to Aristotelian organicism and its associated conception of causality. *Kalam* atomism was quite unlike that of Democritus and other ancient Greek atomists – *kalam* atoms did not persist in time but appeared and disappeared at every instant. There is good reason to assume that *kalam* atomism is derived from Buddhist views of natural philosophy, which also embraced similar ideas of instantaneous-point atoms.

Indeed, atomism occupied a central place in Indian thought, in contrast to its marginal status in the dominant traditions of Greek Platonism and Aristotelianism. Nearly all the major schools of Indian philosophy – Hindu, Buddhist, Jain – adopted atomic conceptions of natural philosophy. When Arab Muslims came to conquer the Persian Sassanian Empire, the views of these Indian schools were actively debated at the centres of learning at Jundishapur – a meeting place for Persian, Greek and Indian scholars. It is also likely that Arabic theological thinkers might have found such atomic views attractive since – unlike Greek atomism, with its close association with atheism – Indian atomism was tied to spiritual traditions in Indian thought. Especially influential must have been Buddhist atomism, which being closely linked with its process thought, saw atoms as appearing and disappearing in accord with the Buddhist law of dependent origination. It is but a natural step to see the law itself as a regulatory principle subject to the will of the omnipotent God of Islam. This makes it likely, as Fakhry has noted, that *kalam* atomism was inspired by Indian rather than Greek ideas:

> With hardly a single exception, the Muslim theologians accepted the atomic view of matter, space, and time, and built upon it an elaborate theological edifice over which God presided as an absolute sovereign. We shall have occasion to consider this atomic theory later, but it is noteworthy that some of its important divergences from Greek antecedents, such as the atomic nature of time, space and accidents, the perishability of atoms and accidents, appear to reflect an Indian influence. The two Buddhist sects of Vaibhashika and Sautrantika, the two Brahmin sects of Nyaya and Vaishashika, as well as the Jaina sect, had evolved by the fifth century an atomic theory, apparently independent of the Greek, in which the atomic character of matter, time and space was set forth and the perishable nature of the world resulting from their composition was emphasized.[3]

Indeed, many other ideas characteristic of *kalam* – the search for union with God, the belief in intellectual illumination as a way to knowledge superior to empirical or rational methods, the cultivation of the symbolic interpretation of religious texts – can also be seen to have parallels in Hindu, Buddhist and Jain traditions. This lends even more credence to the notion that *kalam* atomism has Indian, rather than Greek, origins. But what is significant is that the *kalam* thinkers, especially culminating with Ghazali, adopted the metaphysics of instantaneous-point atoms but maintained that each appearance and disappearance was an isolated event that did not influence any other.

In contrast to the Buddhist notion of the law of dependent origination that regulated these changes, *kalam* held that God was the sole influence on these events and that the world was generated continually *ex nihilo* by God's Will. This even extended to the properties of atoms – accidents, as the *kalam* evocatively described them. Such properties were made to be at their instant of appearance by God's Will alone. Thus the universe is at all times subject to the Will of God and could be made to change at any point in time should God change His mind. It is the habits of God that sustain the order of the universe, and miracles are events that happen when God suspends His habits. By adopting such a metaphysics of nature the *kalam* sought to demonstrate the power of God in sustaining the cosmos, the mercy of God in keeping its regularities and the possibility of miracles documented by scripture as events that happened when God changed his habits.

Ghazali's concept of cause-effect relations is predicated on the ontological atomism of *kalam* and God's role in determining the appearance and disappearance of atoms. In his study *The Incoherence of the Philosophers*, Ghazali articulates why the relation between cause and effect cannot be treated as a necessary linkage:

> According to us, the connection between what is usually believed to be a cause and what is believed to be an effect is not a necessary connection; each of the two things has its own individuality and is not the other, and neither the affirmation nor the negation, neither the existence nor the non-existence of the one is implied in the affirmation, negation, existence, and non-existence of the other – e.g., the satisfaction of thirst does not imply drinking, nor satiety eating, nor burning contact with fire, nor light sunrise, nor decapitation death, nor recovery the drinking of medicine, nor evacuation the taking of a purgative, and so on for all the empirical connections existing in medicine, astronomy, the sciences and the crafts. For the connection in these things is based on a prior power of God to create them in a successive order, though not because this connection is necessary in itself, and cannot be disjoined – on the contrary, it is in God's power to create satiety without eating and decapitation without death and so on with respect to all connections.
>
> The philosophers however deny this possibility and claim that, that is impossible. To investigate all these innumerable connections would take too long, and so we shall choose one single example, namely the burning of cotton through contact with fire; for we regard it as possible that the contact might occur

without the burning taking place, and also cotton might be changed into ashes without any contact with fire, though the philosophers deny this.[4]

Ghazali developed his argument to refute the notion of self-evident first principles as the basis for a science. He agreed that every argument must start with premises that have to be taken for granted and that every science that permits us to deduce cause-effect relations must start with first principles. He also agreed that reason could be one way of arriving at first principles, as Plato held, and that experience can be another way if we follow Aristotle. But he questioned the assumption that such first principles can ever be established as self-evident. He argues as follows. Suppose we succeed in establishing some principles as self-evident or indubitable by deploying reason or observation. We can then use them to infer cause-effect linkages for phenomena in their domain of applicability. But whatever relations we establish will be necessary relations since they follow deductively from self-evident first principles. This means that God could not have made a world that obeys cause-effect relations other than the ones we now observe. But this would violate the omnipotence of God by placing limits on the regularities He could impose on the universe. Hence the first principles cannot be necessary, but only contingent, relations – relations dependent on the Will of God. If so they cannot be self-evident. If they are not self-evident they can be doubted. Therefore, neither reason nor experience can establish self-evident first principles. All the sciences are grounded in principles that could have been otherwise.

In place of the Aristotelian and Greek conception of cause-effect linkages – an event A is regularly connected to an event B because of a necessary link between A and B – Ghazali argues that these linkages exist because God wills it so. The regularities in the causal nexus we find in the world are due, Ghazali maintains, to 'the habits of God'. Being the products of God's Will they could have been otherwise if God had willed it so. Indeed, since the atoms that appear and disappear at every instant are really what make the world, these regularities are continually at the mercy of God. At any instant God could alter the regularities by changing his habits, and what we call miracles are precisely those events that occur when this happens. In short, it is God's habits that make for the regularities in the causal linkages we perceive in the world and not any necessary connections between a cause A and its presumed effect B.

In modern times similar arguments were presented by the philosopher David Hume, who even seems to adopt some of the same examples used by Ghazali to illustrate his position. The historian of science Osman Bakar makes this point:

> Interestingly enough, in his repudiation of causality, Hume presented arguments very similar to those offered by the Asharites [kalam], but without positing the Divine Will as the nexus between two phenomena which the mind conceives as cause and effect. Moreover, some of his examples were the same as those of the Asharites.[5]

However in contrast to the *kalam*, Hume does not see the regularities in the cause-effect relations we find in nature as due to the habits of God, but to habitual ways of organizing knowledge by human beings. What is significant is that both Ghazali and Hume reject the idea that there is a necessary connection between causes and their effects.

Newton seems to espouse a conception of cause and effect as mediated by contingent relations laid down by God when he writes:

> [I]t seems probable to me, that God in the beginning formed matter in solid, massy, hard, impenetrable, moveable particles of such sizes and figures and with such other properties and in such proportion to space, as most conduced to the End for which he formed them. . . . And therefore, that Nature may be lasting, the changes of corporeal things are to be placed only in the various separations and new associations and motions of these permanent particles . . . it seems to me farther, that these particles have not only a *vis inertiae* [inertial force], accompanied with such passive laws of motion as naturally result from that force, but also that they are moved by certain active principles, such as is that of gravity, and that which causes [chemical] fermentation and the cohesion of bodies.[6]

It is clear that the Newtonian view above not only constitutes a break with the Aristotelian notion of necessary connections linking cause-effect relations, but also embraces the anti-Aristotelian conception of atomism conjoined to the notion that the way these atoms interact is dependent on God's Will. In many ways it is closer to Ghazali's ontology of atomism and conception of contingent cause-effect relations than Aristotelian views. But there are also important differences. Newtonian atoms persist over time and obey laws laid down by God. It raises the question of the extent to which Ghazali could have influenced Newton, how such influence could have been transmitted and also what made

Newton's ontology of persistent atoms and persistent laws so different from Ghazali's views. However, before we address these questions, which suggest that Ghazali could have contributed in some positive fashion to modern science and its epistemology, let us examine the prevalent view that he actually had a detrimental influence upon science.

Ghazali and the decline of arabic science: two contrary views

Ghazali is often blamed for the problems associated with advancing a scientific and technological culture in the contemporary Middle East. This view is endorsed not only by critics of Ghazali who mourn the baneful influence of religion on science, but also by those who celebrate his effort to insulate religion against contamination by science. Such interpretations are not surprising because Ghazali did indeed work relentlessly to purge Islamic civilization of the influence of Aristotelian science – he saw it as in conflict with established religious beliefs since it assumed that the universe was eternal by virtue of having no beginning in time and that God's omnipotence was limited since he could not create any universe other than the current universe conforming to self-evident first principles.[7] As a result Pervez Hoodbhoy, in *Islam and Science: Religious Orthodoxy and the Battle for Rationality*, concludes:

> The rumblings against the secular and universalistic character of Hellenistic knowledge started . . . almost from the time of its introduction into the Islamic culture. But the confusion of competing doctrines, lack of familiarity with the techniques of logic and science, and incessant bickering, did not at first allow for a sustained and decisive attack against rationalism. It was not until the theologian Al-Ghazali – a man who Seyyed Hossein Nasr gratefully acknowledges as having 'saved orthodoxy by depressing science' – that a coherent rebuttal of rationalist philosophy was attempted. With perspicacity, scholarship, and singlemindedness, Al-Ghazali worked tirelessly to rid Islamic culture of the foreign intrusion of Greek thought.[8]

Hoodbhoy notes that central to Ghazali's critique of science was the concept of causality, which is 'the cornerstone of scientific thinking'.[9] Since he adopted the *kalam* belief that the universe is continuously created and destroyed by God, instant by instant, the regularities we observe in the universe cannot be the outcome of causal influences of bodies on one another. There is no cause-effect chain in the universe where one event determines another. God is the sole cause of all events since he produces the atoms that make up the universe at one instant of time, causes them to decay at the next instant, and then recreates the universe immediately thereafter. No event influences another; God is the sole influence on all events. Such an account leaves no scope for the laws of cause and effect so integral to any scientific description of the universe. Consequently Hoodbhoy attributes the decline of science in the Muslim world, and the continual resistance to scientific thinking today in many Muslim cultures, directly to this Ghazalian impact. He writes:

> The decline of science in Islamic culture was contemporaneous with the ascendancy of an ossified religiosity, making it harder and harder for secular pursuits to exist. This does not pinpoint the orthodox reaction against science as the single cause. In particular, it does not exclude economic and political factors. But certainly, as the chorus of intolerance and blind fanaticism reached its crescendo, the secular sciences retreated further and further. Finally, when the golden age of the Islamic intellect ended in the fourteenth century, the towering edifice of Islamic science had been reduced to rubble.[10]

Hoodbhoy's conclusion is endorsed by Edward Grant in his study *The Foundations of Modern Science in the Middle Ages*:

> Most Muslim theologians were convinced that on certain crucial issues Greek logic and natural philosophy – especially Aristotle's natural philosophy – were incompatible with the Koran. The greatest issue that divided the Koran from these disciplines was the creation of the world, which is upheld in the Koran, but was denied by Aristotle, for whom the eternity of the world was an essential truth of natural philosophy. Another serious disagreement arose in connection with the concept of secondary causation. In Islamic thought, the term 'philosopher' was often reserved for those who assumed with Aristotle that natural things were capable of causing effects, as when a magnet attracts iron and causes it to move, or when a horse pulls a wagon and is seen as the direct cause of the motion. On this approach, God was not viewed as the immediate cause of every effect. Philosophers believed with Aristotle that natural objects could cause effects in other natural objects because things had natures that enabled them to act on other things and to be acted upon. By contrast, most Muslim theologians believed, on the basis of the Koran, that God caused

everything directly and immediately and that natural things were incapable of acting directly upon other natural things. Although secondary causation is usually assumed in scientific research, most Muslim theologians opposed it, fearing that the study of Greek philosophy and science would make their students hostile to religion.[11]

But George Saliba contests the idea that Ghazali's impact could have been as negative as often portrayed. In his study *Islamic Science and the Making of the European Renaissance* he points out that many significant discoveries that led to advances in science were made in the Islamic civilization long after Ghazali's demise. There was, he noted, no such deep conflict between science and religion in the Islamic world. To perceive such conflict is merely to project back into medieval Islamic civilization the religion-science conflict that characterizes modern science. According to Saliba many of the Arabic scientists who repudiated Greek science did not do this on religious grounds – they did it because they saw errors and shortcomings in the inherited tradition of Greek science. As examples he cites Muhammad ibn Musa's critique of Ptolemy, Razi's objections to Galen in his *Shukak*, and Ibn al-Haitham's *Doubts Against Ptolemy*.[12] Saliba concludes:

> [T]hose who hold Ghazali responsible for the age of decline, they will have to explain the production of tens of scientists, almost in every discipline, who continued to produce scientific texts that were in many ways superior to the texts that were produced before the time of Ghazali. In the case of astronomy, one cannot even compare the sophistication of the post-Ghazali texts with the pre-Ghazali ones, for the former were in fact far superior both in theoretical mathematical sophistication, as was demonstrated by Khafri, as well as in blending observational astronomy with theoretical astronomy, as was exhibited by Ibn al-Shatir.[13]

Saliba traces the decline of Islamic science not to causes within the culture but to factors beyond that made it suffer, first, a relative decline in contrast to developments in Europe and then an absolute decline as European power constricted the space for Islamic science. This change started in the sixteenth century, when the ruling houses of Europe became capable of outbidding the Arabic-Muslim world in funding scientific and technological research because of the wealth generated as a result of the discovery of America. It allowed Europeans to develop science at a pace far more rapid than elsewhere and made the pace of scientific discovery in other cultures seem to slow down by contrast. But initially there was only such relative decline but no absolute decline; the latter began to happen only after Europe became powerful enough to impose its influence on other civilizations.[14] It leads Saliba to reject the notion that Ghazali's critique of the epistemological basis of Greek science caused the decline of science in the Muslim world.

It is evident that two very different accounts of the history of Arabic science are presented by Hoodbhoy and Grant, on the one hand, and Saliba on the other. The former see science as declining after Ghazali, and the latter as reaching even greater heights. It shapes their judgment concerning the impact of Ghazali on Arabic science. However, careful investigation reveals that these two views are not necessarily incompatible once we recognize that Ghazali did not critique science per se, but only Aristotelian science. The science that progressed before him was Aristotelian science in the Islamic world; the science that progressed after him was largely an anti-Aristotelian tradition that arose as a critique of the Greek heritage of science – as Saliba himself notes. Hence there is no inherent conflict between the two positions, any more than there is an incompatibility in claiming that Greek science declined in the modern era, followed by the growth of modern science. Even if it is correct to assume that Ghazali's impact on the growth of Greek science was negative, he may also have had a positive impact on the subsequent growth in the Arabic world of scientific ideas that were critical of Greek science.

From Greek science to modern science: Ghazali's mediating role

This raises the question of how Ghazalian ideas could have been assimilated by European thinkers. It cannot have been direct because Ghazali's *The Incoherence of the Philosophers* was not translated into Latin. However, his views were rebutted by leading medieval scholars such as Maimonides and Averroes whose works were translated into Latin and had widespread readership in medieval Europe. Maimonides' magnum opus, *The Guide to the Perplexed*, identifies twelve propositions shared by the Mutakallemim, as the *kalam* proponents were known. Of these, the first and central claim of the *kalam* is the following:

> The universe, that is, everything contained in it is composed of very small parts [atoms] which are indivisible on account of their smallness; such an atom has no magnitude; but when several atoms combine, the sum has a magnitude and thus forms a body. . . . These atoms, they believe, are not, as was supposed by Epicurus and other Atomists numerically constant; but are created anew whenever it pleases the Creator; their annihilation is therefore not impossible.[15]

Similarly the profound influence of Averroes in medieval Europe would also have served to spread Ghazali's views among European scholars. Averroes' defence of Aristotelianism against Ghazali in his study *The Incoherence of the Incoherence* made Dante refer to him as '*che il gran commento feo*' – 'he who wrote the grand commentary'. Indeed Averroes' call to separate philosophical and religious truth and his defence of Aristotelianism against Ghazali's attack won enthusiastic support in the many new universities in Europe. His defence of rationalism was seen as questioning clerical control of intellectual debate. But his teachings so alarmed the Catholic Church that ecclesiastical authorities in Paris condemned them in the year 1277. Even Thomas Aquinas, who was sent to Paris to counter anti-clerical ideas, came to be suspected of Averroist tendencies – a charge that was to taint his name long after his death.

Over the next three centuries Averroism continued to persist in European academies.[16] It was a significant force in the University of Padua at the time that Copernicus, Galileo and Harvey studied there and could not have been unknown to them. Thus it is hardly likely that European thinkers would be unfamiliar with *kalam* atomic views and Ghazali's critique of Aristotelian conceptions of causality. Moreover, those in the anti-Peripatetic movement, which developed in the seventeenth century against Aristotle and was closely associated with atomic ideas and empiricism, cannot have been ignorant of Ghazali's arguments. Indeed, it would have been difficult for them not to know of his arguments against Aristotelian ideas, since much of Averroes' critique begins by first presenting Ghazali's position in passages that begin with 'al-Ghazali says'.

At the same time the seventeenth-century anti-Peripatetics would also have been aware of Thomas Aquinas's attempt to incorporate Aristotelianism within the framework of Catholic theology by offering a more nuanced position that took into account the conflicting positions of Averroes and al-Ghazali. He does this by reinterpreting Aristotle as simply an empiricist – a view that now underlines the tendency of modern writers to contrast Aristotelian empiricism with Platonic rationalism. This obscures the strong rationalist element in Aristotle that holds a place for self-evident principles that originate every science. Aquinas also introduces the notion of primary and secondary causes in order to naturalize Aristotle within the context of Catholic Europe and its belief in the omnipotence of God. This enables him to argue that the causes identified by Aristotelians in science are secondary causes laid down by God who is the ultimate and primary agent behind all observed events. Since the principles which define cause-effect relations in the world are freely laid down by God but are not themselves compelled upon God, they cannot be justified by appeal to self-evidence. They can only be discovered and justified by observation and experience.

The intermediate position of Aquinas combines the notion of primary (God-based) causes with secondary (creaturely) causes and is described by Carroll as follows:

Contrary to the positions both of the *kalam* theologians and of their opponent, Averroes, Aquinas argues that a doctrine of creation out of nothing, which affirms the radical dependence of all being upon God as its cause, is fully compatible with the discovery of causes in nature. God's omnipotence does not challenge the possibility of real causality for creatures, including that particular causality, free will, which is characteristic of human beings. . . . For Aquinas 'the differing metaphysical levels of primary and secondary causation require us to say that any created effect comes totally and immediately from God as the transcendent primary cause and totally and immediately from the creature as secondary cause'.[17]

We have seen Newton consider that the way to know the regularities and laws God imposed on atomic behaviour is through experience and observation. God could have made nature obey other quite different laws.[18] This suggests that we have to see the changes from the science of Aristotle and its associated conceptions of necessary causes and rejection of atomism to Ghazali's rejection of necessary causes and embrace of atomism and Aquinas's embrace of primary and secondary causes without atomism to Newton's acceptance of primary and secondary causes with atomism as constituting part of an extended debate in which Greco-Arabic science came to be displaced by modern science. To ignore these dialogical exchanges that mediated this scientific and philosophical reorientation is to ignore the contributions that the *kalam* tradition of philosophy, particularly through Ghazali, made to the making of modern philosophy of science.

Notes

1 Sarukkai 2005, 6.
2 In his study *Parallel Developments* the Japanese comparative philosopher and historian Hajime Nakamura writes:

> In the West the two terms [religion and philosophy] have been fairly sharply distinguished from each other, while in Eastern traditions the dividing line is often difficult to discern. If we insist on being too strict in our definitions, we fail to catch many common problems. It is possible that an idea or attitude held by a Western philosopher finds its counterpart not in an Eastern philosopher but in an Eastern religious thinker and vice versa. (Nakamura 1975, 3)

3 Fakhry 2004, 35.
4 Quoted in Bakar 1999, 92. Bakar's quote is taken from de Berg's translation (1954) of Averroes' *Tahafut al-Tahafut*, in E. J. Gibb Memorial Series, new series 19 (London: Luzac, 316–17).
5 Bakar (1999, 101). Bakar adds that this has led certain scholars to assume that Hume must have been acquainted with *kalam* atomism through the Latin translations of Averroes' *Tahafut al-Tahafut* [*The Incoherence of the Incoherence*] and Maimonides' *The Guide to the Perplexed*.
6 Newton 1952, 400–1. Quoted in Kuhn 1957, 260.
7 Indeed Ghazali writes, 'We must therefore reckon as unbelievers both these philosophers themselves and their followers among the Islamic philosophers, such as Ibn-Sina, al-Farabi, and others, in transmitting the philosophy of Aristotle.' See W. Montgomery Watt 1953, 32–3. Also quoted in Hoodbhoy 1991, 106.
8 Hoodbhoy 1991, 104.
9 Ibid., 105.
10 Hoodbhoy 1992, 95–6.
11 Grant 1996, 178.
12 Saliba 2007, 234–5.
13 Ibid., 237.
14 Ibid., 253–4.
15 Maimonides 1956, 120–1.
16 See Goldstein 1980, 125–6.
17 Carroll 2000, 23.
18 See Osler (1994) for more detailed study of these issues.

References

Bakar, Osman (1999), *The History and Philosophy of Islamic Science*. Cambridge: Islamic Texts Society.

Bala, Arun (2006), *The Dialogue of Civilizations in the Birth of Modern Science*. New York: Palgrave Macmillan.

— (2009), 'Eurocentric Roots of the Clash of Civilizations: A Perspective from History of Science', in Rajani Kannepalli Kanth (ed), *The Challenge of Eurocentrism: Global Perspectives, Policy, and Prospects*, New York: Palgrave Macmillan, pp. 9–23.

Carroll, William (2000), 'Creation, Evolution and Thomas Aquinas', *Revue des Questions Scientifiques*, 171(4): 319–47.

Cohen, H. Floris (1994), *The Scientific Revolution: A Historiographical Inquiry*. Chicago: University of Chicago Press.

Fakhry, Majid (2004), *A History of Islamic Philosophy*. New York: Columbia University Press.

Goldstein, Thomas (1980), *Dawn of Modern Science: From the Arabs to Leonardo da Vinci*. Boston: Houghton Mifflin.

Grant, Edward (1996), *The Foundations of Modern Science in the Middle Ages: Their Religious, Institutional, and Intellectual Contexts*. Cambridge: Cambridge University Press.

Hoodbhoy, Pervez (1991), *Islam and Science: Religious Orthodoxy and the Battle for Rationality*. London and Atlantic Highlands, NJ: Zed Books.

Kanth, Rajani Kannepalli (ed) (2009), *The Challenge of Eurocentrism: Global Perspectives, Policy, and Prospects*. New York: Palgrave Macmillan.

Kuhn, Thomas (1957), *The Copernican Revolution: Planetary Astronomy in the Development of Western Thought*. Cambridge, MA: Harvard University Press.

Maimonides, Moses (1956), *The Guide for the Perplexed*. Reprint, New York: Dover.

Nakamura, Hajime (1975), *Parallel Developments: A Comparative History of Ideas*. London: Routledge and Kegan Paul.

Newton, Isaac (1730), *Opticks*. Reprint, New York: Dover, 1952.

Osler, Margaret (1994), *Divine Will and the Mechanical Philosophy: Gassendi and Descartes on Contingency and Necessity in the Created World*. Cambridge: Cambridge University Press.

Saliba, George (2007), *Islamic Science and the Making of the European Renaissance*. Cambridge, MA: MIT Press.

Sarukkai, Sundar (2005), *Indian Philosophy and Philosophy of Science*. New Delhi: Centre for Studies in Civilizations.

Watt, W. Montgomery (1953), *The Faith and Practice of al-Ghazali*. London: George Allen and Unwin.

CHAPTER 8

SOCIOLOGY OF SCIENCE: BLOOR, COLLINS, LATOUR

Martin Kusch

Introduction

It is difficult to present and discuss the work of David Bloor, Harry Collins and Bruno Latour in one short chapter. Over the last 40 years, among them they have published more than 20 books and several hundred papers. Moreover, the work of each one of them can be subdivided into several stages; each one of them has, sometimes successively, sometimes in parallel, pursued a number of different, only loosely connected projects. And they have fought numerous intellectual battles both with each other and with their many critics. Accordingly, I shall not try to summarize their work as a whole. Instead I want to explain and briefly discuss for each of the three authors their perhaps most important, central and controversial idea.

In Bloor's case I shall focus on his classic four-point strong programme in the sociology of knowledge. This passage from Bloor's most influential book, *Knowledge and Social Imagery* (1976/1991), has been attacked dozens and dozens of times by several generations of philosophers and sociologists of science. I here have space to comment only on one recent philosophical critic, Paul Boghossian (2006). As far as Collins is concerned, the obvious candidate for critical scrutiny is the idea of the experimenter's regress presented in its most extensive form in Collins's chef d'oeuvre, *Changing Order* (1985/1992). This idea too has been challenged from many different directions. I am particularly interested in the suggestion, put forward by Benoît Godin and Yves Gingras (2002), that the experimenters' regress is little more than a reinvention of an ancient sceptical argument.

Latour is best known for his controversial metaphysical views. In his early classic *Laboratory Life* (1979), written jointly with Steven Woolgar, Latour proposes that scientific facts are 'constructed' rather than 'discovered' in the laboratory. In later work, for instance, in *We Have Never Been Modern* (1993) and *Pandora's Hope* (1999), he urges that students of science and technology must not assume a ready-made divide between the natural and the social world and that they must give 'agency' not just to humans but also to things. I shall try to explain why I am not convinced.

Bloor and the strong programme

David Bloor (born 1941) is professor emeritus of the University of Edinburgh. He trained in philosophy, mathematics and psychology at Keele, Cambridge and Edinburgh universities. Throughout his career Bloor worked at the Science Studies Unit (SSU) in Edinburgh. Together with Barry Barnes, David Edge, Steven Shapin and others, he made the SSU a leading centre for the sociology of scientific knowledge. In 1974 Bloor formulated the 'strong programme in the sociology of scientific knowledge' in his book *Knowledge and Social Imagery*. Bloor has written extensively about Wittgenstein, the sociology of mathematics, psychology and aerodynamics. He retired in 2007.

In chapter 1 of *Knowledge and Social Imagery*, Bloor takes his starting point from (then) recent work in the social history of science. He mentions, for example, Thomas Kuhn's early study on the influence of developments in water and steam technology on theorizing in thermodynamics (Kuhn 1959); Donald MacKenzie's investigation into the role eugenics played in Francis Galton's concept of the coefficient (MacKenzie 1981) and Paul Forman's analysis of the impact of anti-scientific *Lebensphilosophie* (philosophy of life) upon the Weimar German physics community as it began to develop an acausal quantum theory (Forman 1971). As Bloor sees it, studies such as these best represent what a sociology of scientific knowledge should aim for. Bloor

(1976/1991, 7) offers the following four key tenets as a summary of the methodology involved:

(1) It would be causal, that is, concerned with the conditions which bring about belief or states of knowledge. Naturally there will be other types of causes apart from social ones which will cooperate in bringing about belief.
(2) It would be impartial with respect to truth and falsity, rationality or irrationality, success or failure. Both sides of these dichotomies will require explanation.
(3) It would be symmetrical in its style of explanation. The same types of cause would explain, say, true and false beliefs.
(4) It would be reflexive. In principle its patterns of explanation would have to be applicable to sociology itself. Like the requirement of symmetry, this is a response to the need to seek for general explanations. It is an obvious requirement of principle because otherwise sociology would be a standing refutation of its own theories.

Bloor calls this the strong – rather than the weak – programme in the sociology of knowledge, first, because it is meant to apply to all sciences (not just the humanities) and, second, because it calls for sociological explanations of both true and false beliefs.

One famous historical study in the sociology of scientific knowledge that clearly follows at least the first three tenets is Steven Shapin's paper on phrenology in early-nineteenth-century Edinburgh, 'Homo Phrenologicus' (1979). Shapin begins by noting that anthropologists have identified three kinds of social interests that motivate preliterate societies to gather and sustain knowledge about the natural world: an interest in predicting and controlling events in the natural world, an interest in managing and controlling social forces and hierarchies and an interest in making sense of one's life situation. The first-mentioned interest hardly calls for further comment. But how does an interest in social control relate to knowledge about the natural world? The answer is that people everywhere use knowledge about the natural world to legitimate or challenge social order. It is almost invariably regarded as strong support for a given social arrangement if it can be made out to be 'natural', that is, in accord with the way the (natural) world is.

Shapin argues the same three kind of interests can also be found that sustaining scientific knowledge – phrenological knowledge in early-nineteenth-century Scottish culture for example. Phrenology was developed in late-eighteenth-century Paris by Franz Josef Gall and Caspar Spurzheim. In Edinburgh these ideas were taken up and championed by various members of the rising bourgeoisie, first and foremost by an Edinburgh lawyer named George Crombie. Phrenologists believed that the mind consists of 27 to 35 distinct and innate mental faculties (e.g., amativeness and tune). Each faculty was assumed to be located in a distinct part, or 'organ', of the brain. Moreover, the degree of possession of a given faculty was thought to be correlated with the size of the respective organ. And, since the contours of the cerebral cortex were taken to be followed by the contours of the skull, phrenologists believed that they could 'read off' the skull of a person which faculties he or she possessed and to what degree. Phrenologists believed that the faculties were innate, but they allowed that the environment could have a stimulating or inhibiting effect upon the growth of the brain organs. They also held that social values and feelings were the outcome of an interaction between individuals' innate faculties and the institutions of a particular society.

How then did this theory serve the aforementioned three interests? There can be no doubt that the phrenologists were genuinely curious about the brain as a natural object. They amassed an enormous amount of detailed knowledge about the convolutions of the cortex; they were the first to recognize that the grey and white matter of the brain has distinct functions; and they noticed that the main mass of the brain consists of fibres. They clearly collected as much information about the brain as they could – with their limited means – hoping eventually to be able to explain more and more of the brain's structure and functioning. Thus the interest in prediction and control was obviously important.

As far as the other two interests is concerned, it is important to note that the advocates of phrenology came from bourgeois and petty bourgeois strata in the society. At the time, these strata were moving up in society. Traditional hierarchies and forms of social control were breaking down as commercial interests became more dominant. The economy was rapidly undergoing a shift from a traditional agricultural to a modern industrialist system. This shift weakened the old aristocracy and worked to the advantage of the middle classes. Phrenology was used as an argument in favour of the change. First, it considerably increased the number of mental faculties over the traditional six. An increased number of mental faculties provided a natural argument for a greater diversity of professions and division of labour. Second, the new faculty of 'conscientiousness' explained the new social reality of competition and contest: this was the faculty that allowed people to compare their standing with that of others. And third, phrenology also made sense of the experience of collapsing hierarchies. Traditional philosophy had put a heavy emphasis on the boundary between spirit and body – metaphorically, 'spirit' stood for the governing elite, 'body' or 'hand' for the workforce. Phrenologists stopped short of equating body and mind, but they made the brain the organ of the mind. In other words, phrenological theory was popular among the rising bourgeoisie since it allowed the latter both to feel at home in the changed socioeconomic situation and to argue against the dominance of the old aristocracy.

It is easy to see that the first three tenets of Bloor's programme – causality, impartiality and symmetry in style of reasoning – are key to Shapin's analysis. Shapin's study proposes a causal explanation for the fact that the members of the Edinburgh bourgeoisie tended to favour phrenology over other theories of the mind. The relevant cause was their interest in making sense of their social situation and in changing society in a way that benefits them. Shapin does not say or imply that this social interest was the *only* cause of the belief in phrenology. Indeed, his reference to the role of the interest in prediction and control (of the natural world) and thus to the phrenologists' detailed brain mapping suggests that other causes, for instance the phrenologists' observations about the brain, also were causes of their belief. Furthermore, Shapin's analysis is impartial; he does not attempt to determine which parts of the phrenologists' or the traditional philosophers' theories were true or false, successes or failures. Shapin's mode of investigation is simply blind to these differences. And thus Shapin's style of explanation is also 'symmetrical': the same types of cause explain true and false beliefs. That is, the phrenologists' various social interests explain (in part) why they opted for their theory, both for the parts we now regard as true and for the parts that we now regard as false.

Let us turn to objections and worries. I mention the first worry only because it comes from an influential and important contemporary philosopher, not because I think it has much merit. In a recent book,

Paul Boghossian suggests that according to Bloor's strong programme, 'our epistemic reasons never make *any* contribution whatsoever to the causal explanation of our beliefs, so that the correct explanation is always exclusively in terms of our social interests' (Boghossian 2006, 112). Boghossian offers the first sentence of the tenet of causality as support for this attribution. Maybe Boghossian would have a point if one ignored the rest of Bloor's book, and thus for instance passages as the following: '[Take] . . . the claim that knowledge depended exclusively on social variables such as interests. Such a claim would be absurd . . .' (Bloor 1976/1991, 166). More seriously, Boghossian even ignores, that is, fails to cite, the second sentence of the causality tenet: 'Naturally there will be other types of causes apart from social ones which will cooperate in bringing about belief'. Such other types of causes include processes in the natural world and our perceptions of and reasoning about these processes. Why are these perceptions and acts of (collective) reasoning not 'epistemic reasons' by Boghossian's standards?

Boghossian's second objection addresses the impartiality and the symmetry tenets. Boghossian alleges that Bloor is committed to a postmodern relativism according to which 'there are many radically different, yet, "equally valid" ways of knowing the world, with science just one of them" (2006, 2). Presumably Boghossian reasons as follows: Bloor's strong programme urges us not to take a stance concerning the truth or falsity, rationality or irrationality, of different belief systems, and he thinks that the same types of causes explain both true and false, both rational and irrational, beliefs. Surely this is tantamount to insisting that there is no intellectual difference between modern physics and voodoo, modern biology and magic. It is true that this is a possible reading if one considers the sentences of tenets two and three in isolation from pretty much everything else Bloor has written in *Knowledge and Social Imagery* and elsewhere. But once this wider evidence is used to clarify the meaning of the strong programme, it is obvious that Boghossian's interpretation is off target. For instance in a much-cited paper, written jointly with Barry Barnes, Bloor writes that his 'equivalence [i.e., symmetry] postulate . . . is that all beliefs are on a par with one another with respect to the causes of their credibility. It is not that all beliefs are equally true or equally false . . .' (Bloor 1982, 23). In other words, Bloor holds that in the case of *all* scientific beliefs – true or false – we should seek sociological (partial) explanations for why they seemed credible to their advocates. But this does not mean that Bloor has any tack with the thought that all beliefs are true or all beliefs are false.

A third (and final) worry also concerns the symmetry principle. Boghossian calls this principle 'impossibly vague' since 'we have not been told what it would be for two explanations to invoke or fail to invoke the same "types" of cause' (Boghossian 2006, 115, fn. 4). Similar concerns have been expressed earlier by other critics. Is an explanation in terms of interest the same type of explanation as an explanation in terms of psychoanalytic categories? Is an explanation of a misperception in terms of a malfunctioning visual system the same 'type' of explanation as an explanation of a correct perception in terms of a working visual system? In answer to this worry it has to be conceded that Bloor does not offer a comprehensive categorization of different types of causes. What he does do, however, is provide an open-ended list of types of sociological explanations. Such types of explanations focus on economic, technical and industrial development, training and socialization, hierarchies and authorities, indoctrination, social influences, ideologies and interests (1976/1991, 6, 32–4, 62, 166). With this list in hand, it seems, the vagueness deplored by Boghossian is no great problem. We can now identify which types of causes Bloor has in mind when he says that the strong programmer offers symmetrical explanations of true and false beliefs; both kinds of beliefs are explained – in part – by social phenomena of the types given.

Collins and the experimenters' regress

Harry Collins was born in 1943. He studied sociology at London University, Essex University and Bath University. His PhD is from Bath. From 1973 until 1995 he worked at the University of Bath, during which time he built up the 'Bath School' in science studies. Between 1995 and 2003 he taught at the University of Southhampton. Since 2003 he has been Distinguished Research Professor at Cardiff University, where he runs the Centre for the Study of Knowledge, Expertise and Science (KES). Collins is best known for his work on the sociology of experimentation, especially in contemporary physics, and for his studies of tacit knowledge and expertise.

Here I shall focus on Collins's work on experimentation in cutting-edge science (e.g., Collins 1985/1992). The feature of this phenomenon that Collins finds most important can best be brought out by contrasting it with two other forms of experimentation: the learner's experiment and the normal-science experiment. The obvious feature of most experiments carried out by students as part of their scientific training is that the students are not meant to find out new things about the world. Indeed, the correct result or outcome of the experiment is known from the start and usually already listed in the textbooks. The real purpose of the experiment is for the students to develop the skill needed for getting to the correct result. As long as they do not manage to obtain the textbook results, the students know that they are still making mistakes in how they conduct the experiment.

The situation in most normal-science experimentation is different: here the skill and instrumentation are known and trusted, but the correct result initially is not. It is a mark of normal science (in the familiar Kuhnian sense) that experimenters have confidence in their abilities and instruments. At least they have this confidence as long as they apply these abilities and instruments in areas that the scientific community regards as investigable. As long as no obvious mistakes are made in the application of familiar investigative tools, the scientific community accepts the results of the recognized practitioners' experimentation. The correct result is whatever is the outcome of the proper application of familiar skills and instruments.

The peculiarity of the cutting-edge experiment is that *both* the correct result and the proper skills and instruments are initially unknown. In cutting-edge science the experimenters are initially uncertain about what the result of their experiment should be, and they are uncertain about what kind of experimental set-up would be needed for achieving this result. They thus find themselves in a 'regress': reflection on what the result should be drives them to consider possible experiments, and reflection on possible experiments drives them to consider what the result should be. This is the experimenters' regress.

It is useful to also think through how the experimenters' regress expresses itself in a controversy in cutting-edge science. Assume two groups of scientists, the As and the Bs, disagree over the existence of some natural phenomenon p. The As insist that p does not exist; the Bs are adamant that it does. The Bs claim to have experimental evidence for their belief; they take themselves to have built a *p*-detector that produces compelling readings for belief in the existence of *p*. The *A*s hold that there cannot be a detector of *p*, since *p* does not exist. And the *A*s are convinced that the *B*s must have misunderstood the working of their alleged 'detector'. To make their case, the *A*s construct what they believe to be a replication of the *B*s' instrument. And they find that their replica detects no *p*, that their replica produces no data that compels recognition of the existence of *p*. The *A*s conclude that the so-called *p*-detector is used properly only if it produces *no data in support of p*. The *B*s are not impressed. Since they are convinced both of the existence of *p* and of the proper functioning of their *p*-detector, they are likely to deny that the *A*s have correctly replicated their – that is, the *B*s' – instrument. As the *B*s see it, the *p*-detector is used properly only if it produces *data in support of p*. It is not obvious how such controversy can be quickly brought to an end: The two sides cannot agree on the existence claim since they cannot agree on the instrument and skills, and they cannot agree on the instrument and skills since they cannot agree on the existence claim. What can break the regress?

Collins's answer can best be introduced by focusing briefly on his best-known case study of the experimenters' regress. In 1969 the American physicist Joseph Weber, of the University of Maryland, claimed to have detected 'high-flux gravitational waves'. Weber's detector was a massive aluminium alloy bar; he assumed that the gravitational attraction between the bar's atoms would change when the bar was hit by gravitational waves originating from exploding supernovae, black holes or binary stars. Weber claimed to have detected several gravitational waves every day and simultaneously on bars 1,000 miles apart. Weber's claim were sceptically received since his insistence on *high-flux* gravitational waves contradicted contemporary cosmological theory. The sceptics built replicas of Weber's instrument and reported no data suggesting the existence of high-flux gravitational waves. Weber rejected his opponents' instruments and measurements as insufficiently true to his model. And thus all elements for the experimenters' regress were in place:

> *But what is the correct outcome?* What the correct outcome is depends upon whether there are gravity waves hitting the Earth in detectable quantities. To find this out, we must build a good gravity-wave detector and have a look.

> But we won't know if we have built a good detector until we have tried it and obtained the correct outcome! But we don't know what the correct outcome is until . . . and so on ad infinitum. (Collins 1985/1992, 84)

Of course in practice the experimenters' regress does not go on ad infinitum. When no theoretical considerations and experimental data can compel consent, scientists resort to other, further criteria. Thus, when Collins (1985/1992, 86) interviewed important physicists at the time of the controversy, he found that they made up their minds about Weber's claims on the basis of elements of the following list:

- faith in experimental capabilities and honesty, based on a previous working relationship;
- personality and intelligence of experimenters;
- reputation of running a huge lab;
- whether the scientist worked in industry or academia;
- previous history of failures;
- inside information';
- style and presentation of results;
- psychological approach to experiment;
- size and prestige of university of origin;
- integration into various scientific networks;
- nationality.

Collins's work on the experimenters' regress has met with critique from many quarters. One often-heard critical comment is that the experimenters' regress is no news; it is simply a restatement of underdetermination arguments familiar from the writings of Duhem or Quine. Duhem taught us that the we can never test a single hypothesis in isolation; an experimental test is always a test of whole groups of hypotheses and auxiliary assumptions. It follows that a negative or unexpected experimental outcome can be accounted for either by abandoning the main hypothesis at issue or else by adjusting or ditching various background hypotheses or auxiliary assumptions. Let us call this form of underdetermination holistic. Quine's emphasis is different; he is concerned with 'contrastive' underdetermination (Stanford 2009): This is the claim that for any given set of data which confirms a given theory, there are other theories that are equally well confirmed by this set of data.

It is clear that the experimenters' regress concerns phenomena of underdetermination; the experimental outcomes of both Weber and his opponents taken together underdetermine the choice between the two hypotheses (of high-flux gravitational waves existing or not existing). This looks similar to Quinean contrastive underdetermination. Or, when the experiments of Weber's opponents failed to find data supporting the existence of high-flux gravitational waves, they rejected the hypothesis that there are high-flux gravitational waves. Weber suggested they instead reject auxiliary assumptions about the proper working of their instruments or computer programs. This fits Durkheim's holistic underdetermination. And yet, it seems to me that despite the presence of these familiar forms of underdetermination, there is also something importantly novel about the experimenters' regress. Both Duhem's holistic and Quine's contrastive forms of underdetermination focus on how *the data leaves the choice of theory underdetermined*. This focus coincides with one-half of the experimenters' regress. But the latter has an equally important second half that does not have a direct parallel in Duhem's and Quine's work. According to this second half, *theory determines the selection of data*. In light of his theoretical assumptions about the existence of high-flux gravitational waves, Weber rejected the data of his opponents. And in light of their theoretical assumptions about the non-existence of said waves, his opponents refused to take Weber's data at face value. The original element of the experimenters' regress lies in the way in which it draws our attention to both connections between data and theory: underdetermination in one direction, determination in the other.

The second criticism of Collins that I want to comment on also charges him with reinventing an old wheel. In this case the old wheel is ancient and early-modern scepticism. Benoît Godin and Yves Gingras (2002) find that the experimenters' regress is little more than a restatement of ideas by Montaigne and Sextus Empiricus:

> To judge the appearances that we receive from objects, we would need a judicatory instrument; to verify this instrument, we need a demonstration; to verify this demonstration, an instrument: there we are in a circle. (Montaigne, quoted in Godin and Gingras 2002, 138–9)

> In order for the dispute that has arisen about standards to be decided, one must possess an agreed standard through which we can judge it; and in order

to possess an agreed standard, the dispute about standards must already have been decided. (Sextus Empiricus, quoted in Godin and Gingras 2002, 139)

In the eyes of Godin and Gingras it will not help Collins to insist, in reply, that, contrary to the ancient sceptics, he does believe in the possibility of scientific knowledge. After all, the result of halting the experimenters' regress is knowledge certified by the scientific community. Let me reconstruct Godin's and Gingras' concerns in my own terms. It is natural to think of a sceptical position (say, about the external world) as consisting of four elements: (1) a sceptical *argument* (say, to the effect that we cannot demonstrate that we are not brains-in-vats); (2) a sceptical *presupposition* (if we cannot show that we are not brains-in-vats, then we cannot know that there is an external world); (3) a sceptical *claim* (we cannot know that there is an external world); and (4) a sceptical *survival plan* (we find ways to make do without assumptions about a mind-independent world). Godin's and Gingras's reading of Collins can then be couched in the following terms. Collins's argument is that the experimenters' regress cannot be broken by scientific considerations pertaining to theory and data. His presupposition is that proper scientific knowledge *must be* based upon scientific considerations pertaining to theory and evidence. Collins's sceptical claim is that we are unable to reach scientific knowledge in cutting-edge science. The factors that halt the experimenters' regress belong to the sceptical survival plan; although the experimenters' regress prevents us from obtaining proper knowledge, the social criteria used by scientists to assess the different sides in a scientific controversy lead to some sort of surrogate for scientific knowledge.

Godin and Gingras support their reading of Collins with the following quotation from Collins:

The list of 'non-scientific' reasons that scientists offered for their belief or disbeliefs in the results of Weber's and others' work reveals the lack of an 'objective' criterion of excellence. There is, then, no set of 'scientific criteria' which can establish the validity of findings in the field. The experimenters' regress leads scientists to reach for *other criteria* of quality. (Collins 1985/1992, 87–8; quoted in Godin and Gingras 2002, 149)

If scientists use *non-scientific* reasons for establishing their theories, then surely the result cannot be proper scientific knowledge. Does it not go without saying that scientific knowledge can only be built upon *scientific* reasons?

In order to defend Collins, we need to pay attention to two considerations. First, note that Collins puts 'non-scientific' and 'objective' in inverted commas. This is to indicate, as Collins himself points out in his reply to Godin and Gingras, that he is here 'quoting scientists' usage, not using [his] own words' (Collins 2002, 157). Collins himself is thus *not* committed to the view that the cited 'other criteria of quality' are non-scientific. This brings me to my second consideration. There are two ways of blocking the sceptic's route from sceptical argument to sceptical claim: deny the sceptical argument or reject the sceptical presupposition. In the case at hand, insist that the experimenters' regress *can* be broken by considerations pertaining to theory and scientific data alone or dismiss the view that scientific reasons must be restricted to such considerations. Call the first anti-sceptical strategy direct and the second diagnostic (cf. Williams 1999). My suggestion can then be put in this way. Collins strikes Godin and Gingras as a sceptic because he sets up and defends the sceptical argument; he shows that the experimenters' regress cannot be broken by theory and data alone. But Collins is not a sceptic about scientific knowledge, he is a *diagnostic anti-sceptic*; he argues that we need to widen our conception of which considerations are permitted to count as scientific in cutting-edge science. Considerations to do with the scientist's track-record, position in various networks or standing in the profession are not unscientific. And thus there is, for Collins, no permissible inference from the presence of such considerations to the absence of scientific knowledge. My interpretation is based on the following passage (the *core set* is the 'set of allies and enemies' that are involved in a cutting-edge controversy:

> Core sets funnel all of their competing scientists' ambitions and favoured alliances and produce scientifically certified knowledge at the end. . . . The core set 'launders' all these 'non-scientific' influences and 'non-scientific' debating tactics. It renders them invisible because, when the debate is over, all that is left is the conclusion that one result was replicable and one was not. . . . If one looks very closely, one can see how the outcome of core set debates is affected by these 'socially contingent' factors, but one can also see how the output is nevertheless what will henceforth be proper knowledge. (Collins 1985/1992, 143)

Note bene: 'proper knowledge', not some sceptical surrogate.

Latour's new metaphysics

Bruno Latour (born 1947) studied philosophy and anthropology in Paris. He joined the Centre de sociologie de l'innovation at the École de Mines in Paris in the 1980s and, together with Michel Callon and others, made it a leading place in the sociology of science and technology. In 2006 Latour moved to the Institut d'Études Politiques. He holds the Gabriel Tarde Chair and is associated with the Centre de sociologie des organisations. Latour is best known for his laboratory studies and the formulation of actor-network theory.

In Latour's case it is difficult to speak of *one* most central idea; too many of his ideas tie for this title. Nevertheless, there is a striking theme that runs through all of Latour's writings from the 1970s until the 2010s, and it is this that I shall explain and briefly assess. The theme I have in mind is Latour's insistence that we have to change the most basic categories in terms of which we think about facts, about the border between the natural and the social and about non-human entities.

Latour took the first step of revising metaphysical categories in his joint book with Steven Woolgar, *Laboratory Life* (1979/1986). To understand the argument of the book, we need to know a couple of things about the history of endocrinology (the science of hormones). The thyroid gland controls our metabolism, our growth and maturation. The hormone 'driving' or 'running' our thyroid gland is called thyrotropin. And the hormone that controls the production or release of thyrotropin is the thyrotropin-releasing hormone, or TRH. During the 1960s endocrinologists tried to determine the chemical structure of TRH. The two most important groups involved in the final race were led by Roger Guillemin and Andrew Schally, both working in the United States Both groups made important contributions to the eventual determination of TRH as a peptide with the formula Pyro-Glu-His-Pro-NH_2. And thus Guillemin and Schally were jointly awarded the Nobel Prize for Medicine in 1977.

Latour spent two years in Schally's laboratory as an anthropological observer during the late 1960s, and thus he had the unique opportunity to watch Nobel Prize–quality research in the making. Latour was one of the first practitioners in the field of science studies to insist on such anthropological fieldwork in scientific laboratories, and his model has proven enormously influential. Lab studies have been central to science studies ever since.

Of the many fascinating results reported by Latour and Woolgar, I shall here concentrate on what they tell us about the 'construction of scientific facts'. To understand their view, it is useful to begin with their taxonomy of different types of statements that can be found in laboratory conversations, in lab memos and in (drafts of) scientific papers. Type 1 statements are conjectures or speculations. Type 2 statements are claims that are not yet accepted by the scientific community. Type 3 statements are reports about what other people claim to have found out; for instance, 'The structure of GH.RH *was reported to be X*'. Type 4 are statements of fact; they result when the 'was reported to be' is replaced by a simple 'is'. Finally, type 5 statements also state facts, but they do so with a different 'modality'; whereas type 4 statements offer information that is still in some sense new, type 5 statements express the obvious, something that needs saying only in very special circumstances. Put differently, type 5 statements are normally taken for granted (1979/1986, 79).

Latour's and Woolgar's aim in *Laboratory Life* is the study of how a number of related statements concerning TRH – for instance, 'TRH is Pyro-Glu-His-Pro-NH_2' – moved up from being statements of type 1 to being statements of type 5. Needless to say, such 'transformation of statement type' is a protracted process of experimenting, theorizing, debating and negotiating. Indeed, *Laboratory Life* offers a fascinating narrative of such transformation both within Schally's laboratory and in the discussions in journals and at conferences between the various protagonists of the story. Without going into the details, the following stages of the historical sequence are important to distinguish. First, in the early 1960s Guillemin succeeded in convincing endocrinologists that a characterization of TRH would be possible provided that endocrinologists joined ranks with chemists, developed new bioassays with hundreds of tons of animal brains and invested in expensive new instruments. Second, by 1966 endocrinologists were convinced that pure TRH had been secured and that the tools of analytical chemistry could now be applied to determine its structure. Third, both Guillemin's and Schally's groups' attempts to analyze TRH involved *synthesis*: they tried out various combinations of amino-acids in order find one that behaves similarly to TRH. Fourth, around 1968 such analysis via synthesis had narrowed down the possibilities to about 20. At this stage Schally's team resorted to thin-layer chromatography and mass spectrometry to rule out all but one of the remaining contenders. By September 1969 Schally's in-house chemist was ready to say that TRH is Pyro-Glu-His-Pro-NH$_2$.

For present concerns it is noteworthy that Latour and Woolgar refuse to speak of the episode as ending with the *discovery of the fact* that TRH is Pyro-Glu-His-Pro-NH$_2$. Instead they say that this fact was 'constructed'. Here are their main reasons. (a) To use the terminology of discovery 'would be to convey the misleading impression that the presence of certain objects was a pre-given and that such objects merely awaited the timely revelation of their existence by scientists' (1979/1986, 128–9). (b) TRH "might yet turn out to be an artefact" (ibid., 176). (c) When a statement becomes a type 4 statement, it 'becomes a split entity'. The two parts are words about an object and the object itself. Subsequently, 'an inversion takes place: the object becomes the reason why the statement was formulated in the first place' (ibid., 177). Speaking the language of discovery obscures this process. (d) To use the language of discovery commits one to realism. But realism is just one of many stances that scientists adopt at various times, others being relativism, idealism or scepticism. Our task as practitioners of science studies should be to understand how scientific debates over the existence of objects get resolved, not take part in these debates (ibid., 179). (e) '. . . "reality" cannot be used to explain why a statement becomes a fact, since it is only after it has become a fact that the effect of reality is obtained' (ibid., 180).

That scientific facts are never discovered was Latour's first substantive metaphysical revision. It has not remained his last. At least equally important is his rethinking of the divide between 'nature' and 'subject / society' and the criticism of Bloor's programme that it involves. Latour applauds Bloor's programme insofar as it demands a symmetrical treatment of both true and false beliefs. At the same time, however, Latour complains that Bloor's approach is deeply *asymmetrical* in other respects; in particular, Latour charges the strong programme with being asymmetrical with respect to the categories of the natural and the social, nature and society. As Latour has it, for Bloor nature figures merely as a passive element with no other role or function than that of a canvas upon which different social groups (of scientists) project different interpretations (Latour 1992).

In so doing, Latour contends, Bloor turns out to be simply another (unoriginal) stage in 'modernism', or in the philosophical tradition from

Kant to Heidegger and Derrida. This tradition, as well as modernism in general, is allegedly marked by ever new attempts to distinguish sharply between nature on the one hand and the individual or social subject on the other hand. In insisting on this divide, the mentioned philosophers defend the freedom and agency of the social realm, while restricting the natural world to being passive, ahistorical, and void of moral or social qualities (Latour 1993). The problem with modernism is that it is living a lie; while philosophers and sociologists have been busy trying to 'purify' the realms of the natural and the social, technologists and engineers have been hugely successful in filling our world with 'hybrids', 'quasi-objects' and 'networks', that is, with entities that cannot be clearly classified as either natural or social. Think of the AIDS virus: it 'takes you from sex to the unconscious, then to Africa, tissue cultures, DNA and San Francisco . . .' (Latour 1993, 2).

What can we do to make things better? Revise our metaphysical categories, is Latour's answer. Accordingly, he is happy to speak of the agency, intentionality and moral qualities of all sorts of mechanical artefacts and biological systems (including mechanical door-closers, lactic acid ferment and scallops); he insists that natural entities have 'historicity' just as we do, and he suggests conceiving of altogether new hybrid entities. The historicity theme is central to Latour's re-conceptualization of experiments as 'events'. After Pasteur had experimented on lactic acid ferment and after the academy had accepted his results, the identity of all three, the ferment, Pasteur and the academy, changed forever:

> If Pasteur wins we will find two (partially) new actors on the bottom line: a new yeast and a new Pasteur! (Latour 1999, 124)

> . . . we should be able to say that not only the microbes-for-us-humans changed in the 1850s, but also the microbes-for-themselves. Their counter with Pasteur changed them as well. (ibid., 146)

And here is Latour's central example by means of which he wishes to introduce new 'actors' or 'actants': it wasn't Oswald who killed J. F. Kennedy, and it wasn't a gun. It was a new actor or actant: 'a citizen-gun, a gun-citizen'. Oswald was 'a different person with the gun in [his] hand', 'the gun [was] different with [him] holding it' (ibid., 179–80). Oswald and the gun each partially define each other's identity, and thus neither deserves to be given ontological priority.

Let me turn to an assessment. Like Bloor and Collins, so also Latour has been the target of much criticism – not least by Bloor and Collins (Collins 1992; Bloor 1997). Above, I have tried to defend Bloor and Collins against one or two lines of attack. In Latour's case I cannot help taking up the role of a critic myself. For lack of space I shall confine myself only to the revision of metaphysics in *Laboratory Life*. The reader should be able to see how my comments would apply to the later work.

My overriding difficulty with Latour's new metaphysics is one of sheer intelligibility. When someone denies one of our common-sense certainties – say, when someone denies that one needs a spacecraft to get to the moon – we feel unsure how to respond. One temptation is to doubt the speaker's sanity; other, less uncharitable, moves are to wonder whether our interlocutor is perhaps giving expression to some religious, poetic or magical sentiment or using the words *spacecraft* and *moon* in unusual ways. Now, when I read in Latour that one should never assume of any scientific fact that it has been discovered, I cannot but feel that a central common-sense certainty is being rejected. Or consider the following more recent passage:

> The attribution of tuberculosis and Koch's bacillus to Ramses II should strike us as an anachronism of the same calibre as if we had diagnosed his death as having been caused by a Marxist upheaval or a machine gun or a Wall Street crash. (Latour 2000, 248)

Here I am at a loss as to what the speaker is actually trying to say. In other words, the concept of the (scientific and non-scientific) discovery of facts about the world is so basic and central in my web of beliefs that to give it up is like throwing out some basic mathematical or logical truths. Of course there are scenarios where we have had to do just that – think of the changes in our beliefs forced upon us by the theory of relativity or quantum mechanics – but in these cases most of us can see how and why we were forced to make the appropriate changes. Nothing of the kind is offered in Latour.

But how can Latour himself think that what he is saying makes sense? After all, no one would think that this brilliant and entertaining man is insane! The only charitable interpretation I can think of is that Latour and Woolgar are using the terms *fact* and *discovery* differently from the way I do. Recall how Latour and Woolgar first introduce 'facts': facts are certain kinds of statements, and facts result from processes of splitting and inversion. These claims do not fit with the way in which we use the term *fact* in everyday or scientific life. They are not observations concerning our ordinary concept of fact; they stipulate and define a different concept. That this is the case does not become obvious to Latour and Woolgar since they never systematically reflect on the distance between common-sense intuitions and their new categories.

If I am right about Latour's and Woolgar's concept of fact – a concept of fact implicitly defined by their various claims about it – the next obvious question must be this: Is their concept so related to ours that it can illuminate our practices in using our concept? This is a question for another occasion. At this stage I am content to have made plausible that the only way to save at least some degree of intelligibility for Latour's claims is to suggest that he *stipulates* rather than *discovers* a new metaphysics.

References

Barnes, B. and D. Bloor (1982), 'Relativism, Rationalism and the Sociology of Knowledge', in M. Hollis and S. Lukes (eds), *Rationality and Relativism*. Oxford: Basil Blackwell, pp. 21–47.

Bloor, D. (1976/1991), *Knowledge and Social Imagery*. Chicago: University of Chicago Press.

— (1999), 'Anti-Latour'. *Studies in History and Philosophy of Science*, 30: 81–112.

Boghossian, P. (2006), *Fear of Knowledge: Against Relativism and Constructivism*. Oxford: Oxford University Press.

Collins, H. M. (1985/1992), *Changing Order: Replication and Induction in Scientific Practice*. London: Sage.

— (2002), '"The Experimenters" Regress as Philosophical Sociology'. *Studies in History and Philosophy of Science A*, 33: 149–56.

Forman, P. (1971), 'Weimar Culture, Causality and Quantum Theory, 1918–1927: Adaptation by German Physicists and Mathematicians to a Hostile Intellectual Environment', in R. McCormmach (ed.), *Historical Studies in the Physical Sciences*, vol. 3. Philadelphia: University of Pennsylvania Press, pp. 1–115.

Godin, B. and Y. Gingras (2002), '"The Experimenters" Regress: From Skepticism to Argumentation'. *Studies in History and Philosophy of Science A*, 33: 133–48.

Kuhn, T. (1959), 'Energy Conservation as an Example of Simultaneous Discovery', in M. Clagett (ed.), *Critical Problems in the History of Science*. Madison: University of Wisconsin Press, pp. 321–56.

Latour, B. (1992), 'One More Turn After the Social Turn . . .', in E. McMullin (ed.), *The Social Dimension of Science*. South Bend, IN: University of Notre Dame Press, pp. 272–94.

Latour, B. (1993), *We Have Never Been Modern*, Harlow: Longman.

— (1999), *Pandora's Hope: Essays on the Reality of Science Studies*. Cambridge, MA: Harvard University Press.

— (2000), 'On the Partial Existence of Existing and Non-existing Objects', in L. Daston (ed.), *Biographies of Scientific Objects*. Chicago: University of Chicago Press, pp. 247–69.

Latour, B. and S. Woolgar (1979/1986), *Laboratory Life: The Construction of Scientific Facts*. Princeton, NJ: Princeton University Press.

MacKenzie, D. (1981), *Statistics in Britain, 1865–1930: The Construction of Scientific Knowledge*. Edinburgh: Edinburgh University Press.

Shapin, S. (1979), 'Homo Phrenologicus: Anthropological Perspectives on an Historical Problem', in B. Barnes and S. Shapin (eds), *Natural Order: Historical Studies in Scientific Culture*. Beverly Hills, CA: Sage, pp. 41–71.

Stanford, K. (2009), *Exceeding Our Grasp: Science, History, and the Problem of Unconceived Alternatives*. Oxford: Oxford University Press.

Williams, M. (1999), 'Skepticism', in J. Greco and E. Sosa (eds), *The Blackwell Guide to Epistemology*. Oxford: Blackwell, pp. 35–69.

CHAPTER 9

ONE CANNOT BE JUST A LITTLE BIT REALIST: PUTNAM AND VAN FRAASSEN*

Stathis Psillos

(T)he world is not a product. It's just the world.

Hilary Putnam, 1991

Introduction

Hilary Putnam and Bas C. van Fraassen have been two pivotal figures in the scientific realism debate in the second half of the twentieth century. Their initial perspectives were antithetical – defining an archetypical scientific realist position (Putnam) and a major empiricism-inspired alternative to scientific realism (van Fraassen). But as the years (and the philosophical debates) went on, there have been important lines of convergence in the stances of these two thinkers, mostly motivated by an increasing flirtation with pragmatism and by a growing disdain towards metaphysics.

Putnam's views went through two major turns, in a philosophical journey he aptly described as taking him 'from realism back to realism' (1994, 494). Being an arch-scientific realist in the 1960s and the early 1970s, he moved to a trenchant critique of metaphysical realism and the adoption of a verificationist-'internalist' approach (what he called pragmatic or internal realism), which he upheld roughly until the end of the twentieth century. Then he adopted a direct realist outlook, what he called 'common sense' or 'natural realism', which was based on the denial of at least some of the tenets of his internalist period (e.g., he abandoned a verificationist conception of truth), while at the same time he tried to avoid 'the phantasies of metaphysical realism'.

It is impossible (or almost so) to cover all aspects of Putnam's realist endeavours. I will therefore focus on his changing views about scientific realism.

Van Fraassen occupied a space in the scientific realism debate that was left vacant by Putnam's critique of fictionalism and verificationism – he favoured an agnostic stance towards the ontological commitments of literally understood scientific theories. His positive alternative to realism, constructive empiricism (CE), was meant to be a position suitable for post-positivist empiricists – that is, philosophers who a) take for granted the empiricist dictum that all (substantive) knowledge stems from experience; b) take science seriously (but not uncritically) as the paradigm of rational inquiry; and c) take to heart all criticism of the positivist approach to science. In more recent work, CE has been placed within a broader framework, known as empiricist structuralism – motivated, at least partly, by Putnam's critique of metaphysical realism.

This chapter will discuss these two philosophers' engagement with scientific realism, hopefully in a way that highlights their overlapping trajectories.

From realism to *realism* and back again

The 'no miracles' argument
Putnam (1975, 73) is the author of the most famous argument for scientific realism; it has become known as the 'no miracles argument' (NMA).

> The positive argument for realism is that it is the only philosophy that does not make the success of science a miracle. That terms in mature scientific theories typically refer (this formulation is due to Richard Boyd), that the theories accepted in a mature science are typically approximately true, that the same term can refer to the same thing when it occurs in different theories – these statements are viewed not as necessary truths but as part of the only scientific explanation of the success of science, and hence as part of any adequate description of science and its relations to its objects.

I will not go into the heated discussion about this argument here (see Psillos 1999, ch. 4). Instead, I will make some observations about its role in Putnam's philosophy of science.

Against positivist empiricism

Putnam's NMA is a *positive* argument, which is meant to supplement a *negative* argument – namely, an argument against reductive empiricist or operationalist approaches to scientific theories. Such approaches had once been popular among empiricists, but it was widely accepted in the late 1930s that theories have excess content over whatever can be fully captured in a strict observational language. Yet it was not really until the early 1950s that it became apparent that the 'excess content' that theoretical terms and predicates have is their factual reference: they designate theoretical/unobservable entities. In his writings in the 1960s, Putnam aimed to motivate and defend this view by arguing systematically against verificationist, reductivist and instrumentalist approaches to scientific theories.

Three of his arguments stick out. The first (1962) relates to his attack on the supposed sharp distinction between observational and theoretical terms. The second (1965) relates to what came to be known as Craig's Theorem: for any scientific theory T, T is replaceable by another (axiomatizable) theory Craig(T), which consists of all and only the theorems of T which are formulated in terms of the observational vocabulary. The new theory Craig(T), which replaces the original theory T, is 'functionally equivalent' to T, in that all observational consequences of T also follow from Craig(T).

Putnam mounted a formidable attack on the philosophical significance of Craig's theorem, arguing a) that theoretical terms are meaningful, taking their meaning from the theories in which they feature, and b) that scientists aim to find out about the *unobservable* world and that theoretical terms provide them with the necessary linguistic tools for talking about things they want to talk about.

Putnam's third argument (1963) relates to the role theories play in the confirmation of observational statements. The idea here is that theories are often necessary for the establishment of inductive connections between seemingly unrelated observational statements.

Given this battery of arguments, the negative argument for scientific realism – namely, that its then extant rivals fail patently to account for the role, scope and aim of scientific theories – was hard to resist.

Realism and the success of science

Note that Putnam's argument for realism refers to Richard Boyd. In his widely circulated and discussed, but still unpublished, manuscript 'Realism and Scientific Epistemology', Boyd tied the defence of scientific realism with the best explanation of the fact that scientific methodology has succeeded in producing predictively reliable theories.

Boyd viewed scientific realism as a historical thesis about the 'operation of scientific methodology and the relation between scientific theories and the world' (1971, 12). As such, realism is not a thesis only about current science; it is also a thesis about the historical record of science. It claims that there has been convergence to a truer image of the world, even though past theories have been known to have been mistaken in some respects. This historical dimension is necessary if the truth (or partial truth, or significant truth) of scientific theories is to be admitted as the best explanation of the predictive reliability of methodology. For unless continuity-in-theory-change and convergence are established, past failures of scientific theories will act as defeaters of the view that current science is on the right track. If, however, realism aims to explain a historical truth – namely, that scientific theories have been remarkably successful in the prediction and control of natural phenomena – the defence of scientific realism can only be a posteriori and broadly empirical. This kind of defence of realism was very congenial to Putnam's overall approach in the 1960s, which rejected the claim that there are absolutely a priori truths.

What is scientific realism?

In light of all this, it is no accident that Putnam takes scientific realism to incorporate three theses:

(a) Theoretical terms refer to unobservable entities (REFERENCE);
(b) Theories are (approximately) true (TRUTH); and
(c) There is referential continuity in theory change (CONTINUITY).

Literal reading of theories

(REFERENCE) implies a certain non-verificationist reading of scientific theories – what came to be known as a 'literal or face-value understanding' of theories. Differently put, it implies a non-revisionist semantics for theories: if theories – taken at face value – talk about electrons and the like, they should be taken to do *exactly* this: to refer to electrons and their ilk. But (REFERENCE) also implies a certain *metaphysical* image of the world: as being populated by unobservable entities. This might not be heavy-weight metaphysics, but, in the context in which it was put forward, it carried considerable weight. It made clear (as Feigl had already recognized) that unobservable entities are no less real than observable entities. It honoured the thought that theoretical entities have independent and irreducible existence. (REFERENCE) implies that the subject matter of science is the unobservable world – at least that it is no less the subject matter of science than are the observable entities.

Truth as correspondence

(TRUTH) takes realism beyond (REFERENCE) in asserting that t-entities (at least those referred to by t-terms featuring in true theories) are indeed real – they populate the world. But for both Boyd and Putnam, (TRUTH) implies a certain understanding of *truth* – namely, truth as correspondence. The chief motivation for such a conception of truth was explanationist. Putnam (and Boyd) insisted that truth (along with reference) plays a key explanatory role: it explains the success of action (more particularly, the success of scientific theories and methodology, in the case of science). This insistence is quite prominent in Putnam's writings until the middle 1970s and especially in his John Locke Lectures, delivered in Oxford in 1976 (cf. 1978).

When it comes to scientific theories, Putnam makes this point vividly by claiming that theories are maps of the world and, in particular, that it is best to view them as such 'if we are to explain how they help us to guide our conduct as they do' (1978, 100). To be successful maps, theories must correspond to some part of reality – pretty much like the successful map 'corresponds in an appropriate way to a particular part of the earth'. Making truth (and reference) explanatory concepts does not imply that they have to be explicitly mentioned in every single explanation of a successful action. Nor does it imply that they should feature prominently in an explanation of language understanding. I can understand how to turn the light on, by flipping the switch, without understanding (or even having the concept of) electricity. It does *not*, however, follow that electricity does not causally explain why the light comes on. Similarly for truth: language can be understood by mastering the use of words and expressions (what Putnam calls 'a use-theory of understanding'), and yet a certain theory of truth can be adopted on the basis of offering an explanation of success.

(TRUTH) then has certain metaphysical implications; namely, that scientific theories are answerable to the world and are made true by the world. However, Putnam did not advance anything more than a certain set of general theses about what truth is *not*. One of them is that truth should not be equated with whatever logically follows from accepted scientific theories, even when these theories are empirically adequate and well-confirmed (cf. 1978, 34–5). The point here is not that theories are or tend to be false. Rather, it is that when truth is attributed to the theory, this is a substantive attribution which is meant to imply that the theory is *made* true by the world, which, in its turn, is taken to imply that it is logically possible that an accepted and well-confirmed theory might be false simply because the world might not conform to it. Let's call this view 'the Possibility of Divergence'. It is meant to capture a sense in which the world is independent of theories, beliefs, warrants, epistemic practices, etc.

Convergence

(CONTINUITY) takes scientific realism beyond (REFERENCE) and (TRUTH) by capturing the all-important notion of convergence in theory-change. Here again, Putnam states (CONTINUITY) in semantic terms: a t-term that features in different theories can nonetheless refer to the very same unobservable entity. This kind of thesis is necessary for convergence, since it secures that successor theories might well talk about the very same entities that their abandoned predecessors did, even though the now abandoned theories might have mischaracterized these entities. Putnam thought that the failure of (CONTINUITY) would lead to a disastrous 'meta-induction': 'just as no term used in the science of more than fifty (or whatever) years ago referred, so it will turn out that no term used now (except maybe observational terms, if there any such) refers' (1978, 25). Then, (REFERENCE) and (TRUTH) go by the board, too.

In a number of papers in the early 1970s, Putnam argued against reference-variance based on the so-called causal theory of reference. Its thrust is this: the reference of a t-term *t* is fixed by the existential introduction of a referent – an entity causally responsible for certain effects to which the term *t* refers. As can be easily seen, the causal theory disposes of semantic incommensurability and establishes (CONTINUITY). It also makes available a way to compare theories and to claim that the successor theory is more truthlike than its predecessors. Besides, the causal theory tallies with Putnam's view that the defence of realism is, by and large, an empirical endeavour.

Against verificationism and fictionalism

Putnam's 'no miracles argument' for realism is the culmination of a complex network of arguments and views that aim to render scientific realism – viewed as endorsing a certain combination of positions – the best way to understand science and to explain its empirical successes. To further support this claim, Putnam (1971) pitted fictionalism against verificationism and argued that fictionalism fails to carve a space for genuine doubt over the reality of the entities that are deemed useful fictions.

Verificationism, Putnam (1971, 351ff) argued, superseded fictional-ism as the dominant anti-realist position because it promised to close the gap between the claim (a) that P is false and the claim (b) that everything in experience is *as if P* were actually true – for example, there are no electrons and yet everything in experience is as if there were actually electrons. Fictionalism takes this combination of claims to be logically consistent precisely because it takes theories at face value. But verifica-tionism closes the gap between (a) and (b) by taking a view of meaning that makes it the case that the meaning of P is exhausted (fully cap-tured) by its empirical content (the difference it makes in experience); hence, two statements (or theories) that have exactly the same empiri-cal content are semantically equivalent, no matter how different they appear to be in their theoretical content. Verificationism, then, defies a face-value reading of scientific theories. According to Putnam, veri-ficationism superseded fictionalism in the minds of many philosophers because it was taken to offer an easy and straightforward way out of the *sceptical* challenge.

Putnam claimed that verificationism – and its concomitant anti-scepticism – was the wrong reason to reject fictionalism. He (1971, 352) called a verificationism-based rejection of scepticism 'the worst argu-ment of all'. Verificationism blocks scepticism by denying the Possibility of Divergence. Both realism and fictionalism honour this possibility.

Why then should fictionalism be rejected? Because it does not make sense to have a *merely* fictionalist stance towards a theory that has been accepted and employed in the explanation and prediction of observable phenomena. The fictionalist would typically read the theory literally, would treat the theoretical concepts as indispensable and would accept a theory 'for scientific purposes' but would refrain from commitment to the reality of the entities implied by the theory since she would take it that the theory – though perhaps empirically adequate – is *false*. What possibly could show to a fictionalist that the theory is true? Putnam takes it that the fictionalist would demand a deductive proof of the theory and rightly objects that if this were the golden standard for acceptance as true, no non-trivial observational statements would be accepted as true either. The fictionalist would end up with scepticism. Putnam challenges the fictionalist to draw and motivate a robust distinction between rationally accepting a theory T (but treating its supposed entities as useful fictions) and rationally accepting that T is true. As Putnam (1971, 354) put it, if one rationally accepts a theory for scientific purposes, 'what further reasons could one want before one regarded it as rational to *believe* a theory?' His answer was that these reasons are good enough!

There have been two major reactions to this argument. The first is to accept Putnam's challenge and to try to defend fictionalism by showing that a certain theory T which assumes – if literally understood – the reality of certain entities can be replaced by another theory T' which does not imply commitment to the reality of the 'suspicious entities'. This is, in effect, the strategy followed by instrumentalists on the basis of Craig's theorem. Putnam's incisive critique of Craig-theorem-based instrumentalism in the 1960s blocked the revival of this position in the philosophy of science. But a position akin to this was revived in the philosophy of mathematics by Hartry Field (1980).

The second reaction to Putnam's argument is, not to adopt fictionalism, but to be agnostic. This is the position articulated by van Fraassen (1980). On this view, the collapse of verificationism does not make scientific realism the only rational option. We discuss this view in the following section.

Isn't 'the worst argument of all' not so bad, after all?

Somewhat surprisingly, in the dying years of the 1970s, Putnam came to accept a third way to resist his own argument against fictionalism based on Michael Dummett's resuscitation of verificationism. Modern verificationism, of the form Putnam came to endorse, takes it that truth is not recognition- or evidence-transcendent. Once this view is adopted, it transpires that the Possibility of Divergence noted above is blocked off: there is no logical or conceptual gap between a suitably justified assertion and truth. In his justly famous *Meaning and the Moral Sciences* (1978), Putnam declared: mea culpa.

Blocking scepticism without verificationism

When Putnam (1971) deplored the 'worst argument' for verificationism – namely, that it blocks off scepticism – he rightly felt the need to show how scepticism is blocked if verificationism is abandoned. His argument is captivatingly simple. The sceptical hypothesis (e.g., the brains-in-a-vat hypothesis) is yet another hypothesis alongside the realist one (e.g., that there is a world of ordinary material objects and human beings); hence, we should examine whether, and to what degree, the sceptical hypothesis is confirmed by the relevant evidence. Confirmation, however, requires a specification of the prior probabilities of the competing hypotheses. Hence, we should look at the prior probability of the sceptical hypothesis. Ranking alternative hypotheses according to their initial (or a priori) probabilities should be based on their respective plausibilities; that is, on judgements as to how plausible they are. But the sceptical hypothesis is far less plausible than the realist one; hence, it is much less confirmed than the realist hypothesis.

Putnam never said what exactly goes into the plausibility judgements. He took it to be enough to stress that

> [t]o accept the plausibility ordering is neither to make a judgement of empirical
> fact nor to state a theorem of deductive logic; it is to take a methodological
> stand. (1971, 353)

What exactly is it to 'take a stand'? Here again, Putnam does not say much. But from the quoted passage, it transpires that taking a stand amounts to making a certain commitment to view the world in a certain way, where this commitment is not idiosyncratic but, in a certain sense, constitutive of rationality. As Putnam (1971, 353) put it, it is the stand taken by all rational human beings – vis-à-vis scepticism, at least.

Why isn't this a good enough answer to scepticism on behalf of a realist? Why, that is, couldn't a realist (Putnam himself) leave open the possibility of scepticism (honoured by the Possibility of Divergence) and, at the same time, argue (along the lines Putnam followed) that the sceptical hypothesis is far less credible than the realist one? Why, in yet other words, did Putnam feel the need to go for verificationism instead?

Blocking scepticism with verificationism

Part of the answer, I think, is connected to Putnam's critique of metaphysical realism (MR). Putnam associated a number of doctrines with MR. MR is supposed to entertain all of the following:

> The WORLD is supposed to be *independent* of any particular representation we have of it – indeed, it is held that we might be *unable* to represent the WORLD correctly (e.g., we might all be 'brains in a vat', the metaphysical realist tells us . . . (1978, 125)

> Truth is supposed to be *radically non-epistemic* . . . (1978, 125)

> The world consists of some fixed totality of mind-independent objects. There is exactly one true and complete description of 'the way the world is'. Truth involves some sort of correspondence relation between words or thought signs and external things and sets of things. (1981, 49)

> (T)here is (. . .) a definite Totality of All Real Objects and a fact of the matter as to which properties of those objects are the intrinsic properties and which are, in some sense, perspectival. (1995, 303)

There isn't enough space here to discuss all these doctrines in any detail, nor to explore their connections. What should be noted is that even if it were rejected that there is such a fixed totality of objects and a fixed set of their intrinsic properties, it would still seem possible that we might be unable to represent the world and that the world might be independent of any particular representation we have of it.

This latter possibility – the Possibility of Divergence – captures more than a kernel of truth in Putnam's characterization of MR: a realist proper (call her metaphysical or not, it does not matter) should honour the Possibility of Divergence, at least in domains where it is extremely plausible to say that there is an external fact of the matter as to what is true or false; something outside our thoughts, language, and minds that is responsible for the correctness of what we come to believe and accept. Putnam's chief point against MR is meant to show that honouring this possibility is incoherent.

The model-theoretic argument against metaphysical realism

To show this, Putnam calls us to envisage an ideal theory T of the world; a theory that satisfies all operational and theoretical constraints; that possesses any property that we can imagine or please except objective truth – which is left open. He takes it that for a metaphysical realist T might still (in reality) be false. His argument then is a *reductio*: if we assume that 'ideal T' might still be false, we end up with absurdity. Let me sketch the argument. Suppose that T says that there are infinitely many things in the world. T is consistent (by hypothesis) and it has only infinite models. By the Löwenheim-Skolem theorem, T has a model of every infinite cardinality (greater than or equal to the cardinality of extra-logical symbols of the language of T). Now, pick a model M of T, having the same cardinality as the WORLD. Devise a one-to-one mapping **m** of the individuals of M onto the pieces of the WORLD and map the relations between the individuals of M directly into the WORLD. These mappings generate a satisfaction relation (in Tarski's sense) – call it SAT* – between (sets of) pieces of the WORLD and terms and predicates of the language of 'ideal T' such that T comes out true of the WORLD. (That is, the WORLD is isomorphic to the model M in which T is true.) The ideal theory has been shown to be true of the WORLD. Then how can we claim that the ideal theory might really be false? This, we are told, would be absurd.

Putnam's challenge is that the very notion of a unique interpretation fixed by the world – implicit in a non-epistemic theory of truth – makes no good sense. Actually, Putnam anticipated an objection that a realist (himself a few years back!) would make: a causal theory of reference would show that, and *explain why*, a particular referential scheme for a language L – call it the intended interpretation – is picked out. Putnam's retort was that a causal theory of reference would be of no help to the realist, since the model-theoretic argument can be extended to a word like 'cause': 'cause' can be reinterpreted no less than other words; in each model M, reference M will be defined in terms of cause M. Then, Putnam said,

> unless the word 'cause' is already glued to one definite relation with metaphysical glue, this does not fix a determinate extension for 'refers' at all. (1980, 477)

Stopping the endless dialogue

As Clark Glymour (1982, 177) has nicely put it, Putnam's argument seems like 'a kind of endless dialogue': whenever one says something about what singles out a referential scheme, Putnam says it is insufficient for the job, for what one says 'adds more theory' which may be reinterpreted in countless ways, and hence it is itself referentially indeterminate. Is the dialogue really endless, though? Note that if it is, Putnam's aim to show that MR is incoherent is far from proven. At most, we have a draw or a stand-off. But the dialogue is not really endless, for Putnam's 'more theory' move misinterprets what the realist claims. As David Lewis (1984) has observed, when a realist suggests an extra constraint – call it C-constraint – that fixes the intended referential scheme, what she suggests is that *in order for an interpretation to be intended, it must conform to* C. Then, the real issue is not whether the *theory* of C will come out true under unintended interpretations. Rather, it is what exactly C is and how it operates. In light of this, Lewis's suggestion is that the appeal to causal considerations in fixing the intended referential scheme is not just adding more theory but offering *constraints* to which an interpretation must conform in order to be intended.

Perhaps the best way to understand what these constraints might be has been suggested by G. H. Merrill (1980), who argued that it is questionable that realism conceives of the world merely as *a set of individuals*; that is, as a model-theoretic *universe of discourse*. The world, a realist would say, is a *structured entity*. Its individuals stand in specific relations to one another or to subsets of individuals. In particular, whereas Putnam's assumption is that the language precedes the world and 'structures' it, the realist position is that the world is *already* structured, *independently* of the language. It is then easily seen why the model-theoretic argument fails. For, an interpretation of the language – that is, a referential scheme – either matches the language to the existing structured world or it does not. If it does not, there is a clear-cut case in which even an ideal theory *might* be false. In particular, if the WORLD is a structured domain, in order for Putnam to have his model-theoretic argument, he would have to show that the mappings from a model M of T onto the WORLD are structure-preserving. Yet, it simply is *not* always possible to produce structure-preserving isomorphisms.

Scientific realism without the metaphysics?

This way of neutralizing Putnam's model-theoretic argument rests on an inflation of the metaphysics of realism. Where does all this leave *scientific* realism? In the midst of his conversion to internal or pragmatic realism, as Putnam tended to call his new verificationist position, he published a piece in which he did endorse scientific realism, suitably dissociated from both materialism *and* metaphysical realism (cf. 1982). What then is scientific realism? Prima facie, it still is what it was taken to be by the early Putnam: theoretical entities have irreducible existence; theoretical terms featuring in distinct theories can and do refer to the same entities; there is convergence in the scientific image of the world; and scientific statements can be (and are) *true*. But – there is always a 'but' – truth is now 'correct assertibility in the language we use' (1982, 197). So scientific realism is retained but dressed up in a verificationist garment.

Is this scientific realism enough? Note that even if truth is tied to justification, one should be careful how exactly this tie is effected. Putnam is indeed extremely careful. As he (1983, 162) put it, he looks for 'a realism which recognizes a difference between 'p' and 'I think that p', between being *right*, and *merely thinking one is right* without locating the objectivity in either transcendental correspondence or mere consensus'. Truth is not a property that can be lost – nor does it have a sell-by date. Hence, the verificationist notion that replaces (or captures) truth should be such that it retains this property of truth. The 'correctness' of an assertion is a property that can be lost, especially if it is judged by reference to current standards or consensus (which come and go). Because of this Putnam ties correctness to 'the verdict on which inquiry would ultimately settle' (1982, 200).

Putnam took it that truth should be constitutively linked with justification, for otherwise 'we cannot say what role it [truth] could play in our lives' (1982, 197). So, there is a set of epistemic constraints that a verificationist notion of truth should satisfy. As Nick Jardine (1986, 35) has aptly put it, the needed concept of truth should be neither too 'secular' nor too 'theological'. It should avoid an awkward dependence of truth on the vagaries of our evolving epistemic values, but it should link truth to *some* notion of ideal epistemic justification. But in its attempt to break away from 'secular' notions of truth and to make truth a standing and stable property, verificationism moves towards a 'theological' notion: the justification procedures become so ideal that they lose any intended connection with humanly realizable conditions. In the end, it becomes either 'secular', resulting in an implausible relativism, or else 'theological' and hence not so radically different from a (realist) non-epistemic understanding of truth, according to which truth *outruns* the possibility of (even ideal-limit) justification.

Why should scientific realism incorporate the claim of mind independence (as elaborated by the Possibility of Divergence)? Why, that is, couldn't someone who accepted the reality of unobservable entities without also rendering them mind-independent (in the above sense) be a scientific realist?

A moral that can be drawn from Putnam's early defence of scientific realism is that the success of science – success that realism is meant to explain – is hard won. It is neither trivial nor in any way guaranteed. The heated debate over the pessimistic induction (see Psillos 1999, ch. 5) has driven the point home that if there is continuity in theory change, this has been a considerable achievement, emerging from among a mixture of successes and failures of past scientific theories. A realist non-epistemic conception of truth and in particular the Possibility of Divergence do justice to this hard-won fact of empirical success and convergence. Given that there is no guarantee that science converges to the truth or that whatever scientists come to accept in the ideal limit of inquiry or under suitably ideal epistemic conditions will (have to) be true, the claim that science does get to the truth (based mostly on explanatory considerations of the sort we have already seen) is quite substantive and highly non-trivial. If, on the other hand, the Possibility of Divergence is denied, the explanation of the success of science becomes almost trivial: success is *guaranteed* by a suitably chosen epistemic notion of truth, since – ultimately – science will reach a point in which it will make no sense to worry whether there is a possible gap between the way the world is described by scientific theories and the way the world is.

It is wrong to pose to the realist the following dilemma: either the concept of truth should be such that cognitive success is guaranteed or else any cognitive success is a matter of pure luck. What Putnam has taught us, to be sure, is that the success of the realist project requires some epistemic luck: if the world were not mappable, science would not succeed in mapping it. But the realist has a story to tell us as to why and how cognitive success, though fortunate and *not* a priori guaranteed, is not *merely* lucky or a matter of chance. The realist story (cf. Psillos 1999) will have to be phrased in terms of the reliability of scientific method and its defence. But there is good reason to think that this story is both sensible and credible.

Constructive empiricism

CE, as characterized by van Fraassen, is a mixture of two theses: an *axiological* and a *doxastic*.

(i)   Science aims at empirically adequate theories; and
(ii) acceptance of scientific theories involves belief only in their empirical adequacy (though acceptance involves *more* than belief: namely, commitment to a theory).

As such, it is contrasted to an analogous doublet of realist theses:

(i')   the aim of science is true theories; and
(ii') acceptance of theories implies belief in their truth.

Given the discussion we have had so far, this is a non-standard way to view scientific realism. But this is not accidental. Van Fraassen approached both scientific realism and constructive empiricism as ways to view science; that is, as ways to view a particular activity or game. This is most naturally understood in terms of its aim (what is the aim of science conceived of as an activity or game?) and of what it is to be counted as success in it (what is involved in meeting the aim of science?). CE, van Fraassen insists, is not an epistemology but a philosophy of science.

If science is seen as an activity, and if SR and CE are seen as rival accounts of this activity, what exactly is the issue between them? For realists, the key issue was the explanation of the (empirically certified and hard-won) success of science and the implication for the epistemology of science that this explanation should have. Not quite so for van Fraassen. Seen as rival accounts of an activity, SR and CE are compared vis-à-vis their ability to explain or accommodate the main actual features of this activity (success being, if anything, just one of them). The explanandum is, we may say, the phenomenology of scientific practice, which, to be sure, should not include the intentions and doxastic attitudes of individual scientists but, instead, the salient features of the activity they are engaged in. The question then is, are there salient features of science (qua an activity or game) that force upon us philosophers of science SR? Or is CE a viable option, too?

For van Fraassen, CE offers an alternative-to-SR philosophical image of science: it views science as an activity or practice which is intelligible and successful, without also accepting that science aims at, and succeeds in, delivering truth. He suggests that it is precisely *this* image – CE – that modern empiricism should juxtapose to scientific realism. The crucial issue, of course, is the grounds on which CE is to be preferred to SR. As we shall see later, van Fraassen has shifted his position somewhat here. The tone of his *Scientific Image* was that CE is rationally superior to SR. Later on, and after a shift in his conception of rationality, it transpired that CE is a rationally permitted view of science.

CE, it should be stressed, is markedly different from old empiricist-instrumentalist positions. Unlike traditional instrumentalism, CE agrees with realism that theories and their concomitant theoretical commitments in science are ineliminable. Van Fraassen accepts that scientific theories should be taken at face value and be understood literally as purporting to describe the reality behind the phenomena:

> [I]f a theory says that something exists, then a literal construal may elaborate on what this something is, but will not remove the implication of existence. (1980, 11)

So theoretical terms should not be understood as useful shorthands for complicated connections among observables. Rather, they should be taken to refer to unobservable entities in the world. Scientific theories are taken as constructions with truth values: true, when the world is the way the theory says it is, and false otherwise (cf. 1980, 10, 38). Unlike Putnam, van Fraassen downplayed the usefulness of the philosophy of language to the philosophy of science. He thought that nothing much could be gained by analysing the language of theories or by looking into the issue of the meaning of theoretical terms (cf. van Fraassen 1980, 56).

Part of the reason for this is that van Fraassen adopted (in fact, co-introduced) the so-called semantic view of theories. According to van Fraassen, theories represent the world *correctly* by having it (the world) as one of their models. But CE does not require that the theories get the world right. Rather, it requires that theories be empirically adequate. This idea of empirical adequacy was meant to capture the old instrumentalist conception that theories should aim to save the phenomena. But whereas the traditional conception, bound as it was to the syntactic view of theories, took it that a theory is empirically adequate (i.e., it saves the phenomena) if and only if all of its observational consequences are true, van Fraassen cast this requirement in model-theoretic terms: for a theory to be empirically adequate it should be the case that the structure of the phenomena is embedded in one of the models of the theory (i.e., that the structure of the phenomena is isomorphic to an empirical substructure of a model of a theory). This way of casting the requirement of empirical adequacy frees it from the commitment to a distinction between observational and theoretical *vocabulary*. At the same time, it honours the instrumentalist (and, as noted above, fictionalist) claim that a theory may be empirically adequate and false: a theory may save all observable phenomena and yet fail to correctly describe their unobservable causes. However, if a theory is solely about observable entities, empirical adequacy and truth coincide.

In recent work, van Fraassen has defended CE as a species of structural empiricism (or empiricist structuralism, as he prefers to put it). This is a view about science and not a view about nature, as van Fraassen (2008, 239) is fond of saying. It incorporates the following two theses:

I. Science represents the empirical phenomena as embeddable in certain *abstract structures* (theoretical models).
II. Those abstract structures are describable only up to structural isomorphism.

Being empiricist, this position is focused on observable phenomena. These are taken to be the sole target of scientific representation. The means or the vehicles of representation are theoretical models – qua abstract mathematical structures – but precisely because mathematical structures can represent only up to isomorphism, the phenomena are described – through science – only up to isomorphism. So all we can know – through science anyway – is structure. This, to be sure, is fully consistent with the thought that we know a lot about observable entities – which knowledge is not structural, since it is supposed to be theory-independent. We shall discuss later some of the problems that empiricist structuralism faces – problems which are intimately connected with those Putnam thought metaphysical realism faces.

Against metaphysics

Van Fraassen has taken it to be a merit of his own empiricism that it delivers us from metaphysics. Realism, van Fraassen says, buys into inflationary metaphysics. What is certainly right is that realist views aim to explain certain phenomena (broadly understood) by positing entities which are said to be causally-nomologically responsible for the explananda. Let's call this realist stance 'explanation by postulation'. To some empiricists (notably Duhem and van Fraassen), the critique of metaphysics is tied to the critique of explanation by postulation. This is supposed to be the pinnacle of inflationary metaphysics.

There are at least two distinct ways in which scientific realism can buy into metaphysics. The first is to adopt the view that the world has a deep and, by and large, unobservable structure – it is made up of entities and causal-nomological relations among them – which is constitutive of and causally responsible for the behaviour of the observable entities. Why, one may wonder, is this kind of explanation-by-postulation inflationary? In a sense, it obviously is: it proceeds by positing further entities that are meant to explain the life-world and its (typically non-strict) laws. But in another sense, it isn't. For if you think of it, it proceeds by positing micro-constituents of macro-objects, whose main difference from them is that they are, typically, unobservable. That a putative entity is unobservable is, if anything, a relational property of this entity and has to do with the presence of observers with certain sensory modalities (of the kind people have) and not others. No interesting metaphysical conclusions follow from this fact, nor any seriously controversial ontological inflation.

The other way in which scientific realism can go into metaphysics is to adopt a certain neo-Aristotelian conception of the deep structure of the world; in particular, one that posits 'regularity enforcers' (e.g., powers). Indeed, an increasing number of realists wed realism with neo-Aristotelianism (cf. Sankey 2008). But, I think (cf. Psillos 2011a), this isn't mandatory for realism.

Van Fraassen's critique of realism has been based, at least occasionally, on running together these two ways to link scientific realism to metaphysics. Consider the following arguments:

> From the medieval debates, we recall the nominalist response that the basic regularities are merely brute regularities, and have no explanation. So here the antirealist must similarly say: that the observable phenomena exhibit these regularities, because of which they fit the theory, is merely a brute fact, and may or may not have an explanation in terms of unobservable facts 'behind the phenomena' – it really does not matter to the goodness of the theory, nor to our understanding of the world. (1980, 24)

> The realist asks us to choose between different hypotheses that explain the regularities in certain ways; but his opponent always wishes to choose among hypotheses of the form 'theory T_i is empirically adequate'. So the realist will need his special extra premiss that every universal regularity in nature needs an explanation, before the rule will make realists of us all. And that is just the premiss that distinguishes the realist from his opponents. (1980, 21)

A number of points can be made against them. First, from the (suspicious anyway) claim that every regularity needs an explanation, it does not follow that it needs a non-regularity-based explanation. A realist can be happy with the thought that it is regularities all the way down; that is, that less fundamental (observable) regularities are explained by more fundamental (framed in terms of unobservables) regularities. Scientific realism does not have to explain the regularity there is in the world by positing regularity enforcers – that is, entities of distinct metaphysical kind that ground , hence explain, the regularities. Second, the claim that a theory is empirically adequate is already 'inflated' vis-à-vis the available data, which show at most that a theory is unrefuted. This could be taken as a brute fact. The very idea that this fact could be explained by the claim that the theory is empirically adequate shows that even the constructive empiricist does not stomach brute facts all too easily.

Concerning this last point, van Fraassen fully grants that going for empirical adequacy involves making a claim that goes well beyond the available data. He nonetheless takes it that

> there is a difference: the assertion of empirical adequacy is a great deal weaker than the assertion of truth, and the restraint to acceptance delivers us from metaphysics. (1980, 69)

But this is an illusion. It does not deliver us from metaphysics. At the very best, it gets away with less metaphysics!

What then is van Fraassen's case *against* scientific realism? Ultimately, it is that the advocates of SR make a 'leap of faith', which is 'not *dictated* by reason and evidence'. Well, this is fine; especially if we read this 'dictation' as we should – namely, that scientific realism (or the truth of a theory) is *proved* by reason and/or evidence. No such proof is, or ever was, forthcoming. We have already seen Putnam claiming that some judgements are neither the product of empirical evidence nor of logic, but amount to 'taking a stand': to viewing the world in a certain way. Putnam, you may recall, took this stand-taking to be constitutive of rationality, since the latter isn't exhausted by the dipole: empirical evidence and logic.

In his writings from the late 1980s on, van Fraassen has developed a 'new conception of rationality' – foreshadowed in the passage above – in view of which, even by his own lights, empiricism and realism are compatible. According to it,

> what it is rational to believe includes anything that one is not rationally compelled to disbelieve. And (. . .) the rational ways to change your opinion include any that remain within the bounds of rationality. (1989, 172–3)

It follows that though it is rational to form beliefs that go beyond the evidence, these beliefs are not rationally *compelling* by virtue of substantive principles and ampliative rules. It can be easily seen that one can be a scientific realist and adopt van Fraassen's conception of rationality: belief in electrons and the like may well come out rational under van Fraassen's conception of rationality, since it is not rationally forbidden. But so may disbelief in them (or agnosticism about them). Hence, van Fraassen's conception of rationality is suitable for constructive empiricists in that it shows that belief solely in the empirical adequacy of theories is rational.

The point is not that CE is irrational – of course it isn't! Rather, it is that a) van Fraassen's conception of rationality is too liberal and b) there is still room for comparative judgement of rationality: some positions are more rational to occupy than others (cf. Darwinism and creationism). I have criticized van Fraassen's new conception of rationality elsewhere (see Psillos 2007).

Philosophical therapy?

As noted already, scientific realism does avoid Putnam's model-theoretic argument by buying into a certain metaphysics: a way of viewing the world as having a certain determinate natural structure.

This is clearly something van Fraassen is not happy with. But his own empiricist structuralism falls prey to Putnam's model-theoretic argument as well. For if an ideal theory cannot fail to be false – if, you may recall, there is always some satisfaction relation, SAT*, between (sets of) pieces of the WORLD and terms and predicates of the language of 'ideal T' such that T comes out true of the WORLD – the very gap between empirical adequacy and truth collapses. So here is a dilemma: either CE has to buy into some substantive metaphysics (one that allows that there is a difference between empirical adequacy and truth), or CE collapses into realism. Note that it is not an option for CE to claim that SAT* is not intended, because though the intended interpretation of the language is not fixed by the world, it is still fixed by the intentions and practices of the language users. This option, with which Putnam's internal realism flirted if it did not directly adopt it, is no less metaphysical than the straightforward realist enough. As van Fraassen (1997, 38) rightly notes, this option imputes unprecedented metaphysical powers to persons.

Van Fraassen claims that Putnam's model-theoretic argument can be dissolved, without adopting any suspicious looking metaphysical postulates. His central idea is captivatingly simple. Putnam, we have seen, equates truth with truth-in-an-interpretation, and since the latter is always available for a consistent theory, so is the former. Not so fast, van Fraassen replies. We know already that the equation between truth and truth-in-some-interpretation is illegitimate when it comes to our own language, the language we understand and use to make contact with the world. For our own language, we are not at liberty to pick any interpretation whatever. The interpretation is already fixed, as it were.

But which language *is* our own? This, van Fraassen says, is an indexical matter; it is the language we actually use and understand. If we lose sight of this pragmatic dimension of language use, van Fraassen (1997, 21) adds, we are tempted to think that the gap between truth and truth-in-an-interpretation 'might be filled by metaphysics'. But if we keep an eye on this dimension, there is no gap to be filled at all for our own language.

There is an obvious retort to this line of reasoning. Could we not consistently think that we might be wrong in interpreting our language the way we do? Could it not be the case that the extensions we assign to predicates in our language do not cut the world at its joints? If these are genuine possibilities, isn't there an opening for metaphysics to get in by virtue of the claim that it is the world that ultimately determines the *correct* interpretation of the language? This, you might recall, is in essence, Lewis's answer to Putnam's model-theoretic argument.

Van Fraassen's reply is most instructive. As I understand it, it goes like this: though it is indeed possible that our own language might have the wrong interpretation (it might fail to cut the world at its joints, as it were), we cannot (in our own language) coherently deny that it (the interpretation-in-use, as it were) is the right interpretation. This situation is supposed to be an instance of what van Fraassen calls pragmatic tautologies: propositions that can be false (i.e., they are not necessarily true) and yet are such that they cannot be coherently denied. Take the following two statements:

(A) *X* believes that P, but it is not the case that P.
(B) I believe that P, but it is not the case that P.

The person *X* in (A) might be myself. Hence, the content of (A) and (B) might well be the same. This shows that (B) is nothing like a formal contradiction. And yet (B) cannot be coherently asserted by me – it is a (pragmatic) contradiction. Van Fraassen's claim is that statements that fix the reference in one's own language (e.g., *cat* refers to cats) are pragmatic tautologies in one's own language: they cannot be coherently denied. As such, he thinks, they raise no metaphysical anxieties about how reference is fixed and what role the world might have in reference-fixing. When further metaphysical worries are raised, van Fraassen (1997, 39) says, they should be treated as needing 'philosophical therapy'.

I am not so sure. To fix our ideas let us make statements (A) and (B) more concrete:

(A') *X* believes that *X*'s language cuts the world at its joints, but it is not the case that *X*'s language cuts the world at its joints.
(B') I believe that my language cuts the world at its joints, but it is not the case that my language cuts the world at its joints.

(B') is supposed to be pragmatically incoherent, while (A') is not. But there is traffic between the two, as when I am in a reflective mode – when I treat myself as 'he', as it were. I know that (B') *might* be true, and I know that I can assert that it *is* actually true if I take the third-person perspective on myself – that is, to assert (A'). So I can raise the question whether *my* language cuts the world at its joints (equivalently, how is the reference of the linguistic items of *my* language fixed?), and I can at least attempt to answer it by letting the world do most of the work in answering it. That is, I can coherently let the world do most of the work in making the case that *natural* classes are the extensions of the predicates of my language. If indeed there is an objective criterion of rightness when it comes to reference-fixing – and both van Fraassen and Putnam think there is – this criterion has to hold for my language, too. But then it seems van Fraassen cannot so easily escape from metaphysics by going for the supposedly harmless pragmatic tautologies.

Ultimately, van Fraassen has had to turn against his former self and abandon the realist background that accompanied CE – in particular, the thought that truth is correspondence with reality. He says this much explicitly in his more recent work.

> Once again we find ourselves with an idea akin to, of a piece with, the correspondence theory of truth, the idea that there is a user-independent relationship between words and things that determines whether a sentence is true or false. Such an idea cannot be carried through without postulating a good deal of ontological flora and fauna beyond concrete individuals. But we have discussed this issue sufficiently above, we don't need to repeat the argument against such presuppositions (2008, 252).

His thought is of a piece with the middle Putnam of internal realism, at least in so far as making the notion of truth-as-correspondence the culprit. But in the case of van Fraassen, we are not told with what to replace it.

Empiricism has always been anti-metaphysics. Yet, logical empiricism – which van Fraassen has attacked – aimed to occupy a position such that the critique of metaphysics left the world as described by science intact (see Psillos 2011b). Van Fraassen's CE has taken it to be the case that the world as described by science was too metaphysical for the empiricist to feel at home in it. Hence, CE marked a revisionist stance towards science, a stance according to which belief in unobservables – and not just in a theory-free standpoint to view the world – was optional. Traditional empiricism took it that the critique of metaphysics was tied to a critique of language and of the limits of meaningful discourse. Van Fraassen disdained verificationism and its dependence on language. The irony is that the rescue from metaphysics comes again from, and through, language, though with a pragmatic account of it.

Concluding thoughts

Complaining against the criticism that his middle views flirted with idealism, Putnam (1994, 462) noted that he never denied that our practices were 'world-involving'. Van Fraassen, too, never doubted that the rightness of opinion depends on what the world is like (cf. 1989, 177). But for the world to be involved in any way whatever with our practices or for the world to be a certain way and not another, it is required that the world – even if understood in a relatively minimal fashion as whatever resists our theorizing – must have *some* structure; it must be, to some extent at least, ready made. Both Putnam and van Fraassen have resisted this image of the 'ready made' world. They have both seen this image as overly metaphysical. Their resistance to realism has been motivated, at least to a considerable extent, by the thought that realism is wedded to inflationary metaphysics. And their recoil from realism has been motivated, at least partially, by the thought that 'proper' views of language and science should deliver us from metaphysics. It is still an open question – to me at least – exactly how much metaphysics (scientific) realism requires or implies, apart from whatever commitments are necessary for securing the Possibility of Divergence. If this is thin metaphysics, so be it!

Note

* A paraphrase of Christopher Hitchens's 'One cannot be just a little bit heretical', from Hitch-22: A Memoir (419).

References

Boyd, Richard (1971), 'Realism and Scientific Epistemology' (unpublished typescript).

Glymour, Clark (1982), 'Conceptual Scheming or Confessions of a Metaphysical Realist', *Synthese*, 51: 169–80.

Jardine, Nick (1986), *The Fortunes of Inquiry*. Oxford: Clarendon Press.

Lewis, David (1984), 'Putnam's Paradox', *Australasian Journal of Philosophy*, 62: 221–36.

Merrill, G. H. (1980), 'The Model-Theoretic Argument Against Realism', *Philosophy of Science*, 47: 69–81.

Psillos, Stathis (1999), *Scientific Realism: How Science Tracks Truth*. London: Routledge.

— (2007), 'Putting a Bridle on Irrationality: An Appraisal of van Fraassen's New Epistemology', in B. Monton (ed.), *Images of Empiricism*. Oxford University Press, pp. 134–64.

— (2011a), 'Scientific Realism with a Humean Face', in Juha Saatsi and Steven French (eds), *The Continuum Companion to the Philosophy of Science*. London: Continuum, pp. 75–95.

— (2011b), 'Choosing the Realist Framework', *Synthese*, 190: 301–16.

Putnam, Hilary (1962), 'What Theories Are Not', in E. Nagel, P. Suppes and A Tarski (eds), *Logic, Methodology, and Philosophy of Science*. Palo Alto, CA: Stanford University Press. Reprinted in Putnam (1975), *Mathematics, Matter and Method*.

— (1963), '"Degree of Confirmation" and Inductive Logic', in P. Schilpp (ed.), *The Philosophy of Rudolf Carnap*. La Salle, IL: Open Court. Reprinted in Putnam (1975), *Mathematics, Matter and Method*.

— (1965), 'Craig's Theorem', *Journal of Philosophy*, 62: 250–60. Reprinted in Putnam (1975), *Mathematics, Matter and Method*.

— (1971), *Philosophy of Logic*. London: George Allen and Unwin.

— (1975), *Mathematics, Matter and Method*, Philosophical Papers, vol. 1. Cambridge: Cambridge University Press.

— (1978), *Meaning and the Moral Sciences*. London: Routledge and Kegan Paul.

— (1980), 'Models and Reality', *Journal of Symbolic Logic*, 45: 464–82.

— (1981), *Reason, Truth and History*. Cambridge: Cambridge University Press.

— (1982), 'Three Kinds of Scientific Realism', *Philosophical Quarterly*, 32: 195–200.

— (1983), *Realism and Reason*, Philosophical Papers, vol. 3. Cambridge: Cambridge University Press.

— (1994), 'Sense, Nonsense and the Senses: An Inquiry into the Powers of the Human Mind', *Journal of Philosophy*, 91: 445–517.

— (1995), *Words and Life*. Cambridge, MA: Harvard University Press.

Sankey, Howard (2008), *Scientific Realism and the Rationality of Science*. Aldershot: Ashgate.

van Fraassen, Bas C. (1980), *The Scientific Image*. Oxford: Clarendon Press.

— (1989), *Laws and Symmetry*. Oxford: Clarendon Press.

— (1997), 'Putnam's Paradox: Metaphysical Realism Revamped and Evaded', *Philosophical Perspectives*, 11: 17–42.

— (2008), *Scientific Representation: Paradoxes of Perspective*. Oxford: Clarendon Press.

CHAPTER 10

BEYOND THEORIES: CARTWRIGHT AND HACKING

William Seager

The realism issue

Scientific realism (SR) is a claim about the metaphysical aims of natural science: the goal of science is truth. To most readers the truth of SR must surely border on the obvious, if not the trivial. Nonetheless, there are several powerful arguments which cast doubt on the viability of SR. Since these issues are substantially covered elsewhere in this volume, only a brief discussion is required here. The first issue is less an argument than an influential remnant of logical positivism. Although the positivists were extremely pro-science, they imposed strictures on its methodology and semantics that made SR hard to sustain. The methodological restriction demanded that science deal only with objective, inter-subjectively available data. The semantic restriction attempted to impose the condition that, except for *analytic* truths, meaningfulness depended on empirical verifiability. While the explicit aim of logical positivism was to destroy speculative metaphysics, it can obviously be wielded against scientific claims that at least appear to transcend the realm of objective data; for example, in their appeal to unobservable, theoretical entities such as electrons, quarks, germs and on the like.

A second anti-realist argument aims to provide some real content to the demand that belief in scientific doctrine should not transcend the empirical. This argument depends upon the so-called pessimistic induction (see Laudan 1981). The history of science is a history of failure; the past four centuries is littered with discarded theories which, after more or less successful runs of various lengths, eventually succumbed to experimental refutation. The pessimistic induction is the claim that, very probably, all the theories we possess are false. Note that the induction fails if we take empirical adequacy as our standard of scientific achievement for, unlike truth, empirical adequacy comes in quantifiable degrees. Even as theories fall by the wayside they retain whatever level of empirical accuracy they had managed to attain.[1]

Still, one might hope that eventually we will somehow discover a true theory so that at least the general goal of SR can be maintained. The third anti-realist argument concludes that this is a vain hope. Usually called the argument from underdetermination or redundancy, the key idea here is that for any given theory, T, there is a large, perhaps infinite, set of possible alternative theories which make exactly the same empirical predictions as T. This is also an old argument, whose origin lies in the writings of Pierre Duhem (Duhem 1914/1991).

One reply is that although there may be, in principle, a plethora of alternative theories this point is philosophically sterile. In the first place, due to the sheer difficulty of theory formulation, we never actually possess more than a very few, if any, serious alternatives. In the second place, the alternatives will differ in their intrinsic plausibility.

The anti-realist will hardly be moved. If our care is for truth, what could be the significance of the peculiar limitations of human creativity and our parochial judgments of plausibility? Plus, of course, the anti-realist is not asking us to renounce the empirical gains of science but only the optional accessory of belief in the non-empirical parts of scientific theory. Thus any judgment of the plausibility of a theory would seem to entail that greater plausibility be assigned to the restricted belief that the theory is empirically adequate.

Can the realists advance positive arguments? There are two main arguments: the 'no miracles' argument and a broad appeal to abduction or inference for the best explanation (IBE).

The basic claim of the former is that unless our theories are (more or less) true, there empirical success would be miraculous. Consider a famous example: Dirac's prediction of antimatter. Dirac first developed the relativistic wave equation for the electron and then noticed that there were two distinct classes of solutions, apparently corresponding to two sorts of particles, one of which turned out to be the positron. How is such success possible if the theory is not getting at the truth?

But, in light of the pessimistic induction, we find that most scientific theories have turned out to be false yet were capable of considerable empirical success and frequently made novel predictions. The phlogiston theory of combustion was able to predict that metals could be reconstituted from their combustion products – what we call the metals' oxides (see Carrier 2004).

Of course, that false theories can possess these virtues does not answer the question of why such theories enjoy empirical success. One anti-realist answer is that we have, so to speak, selectively bred our theories precisely for empirical success (van Fraassen 1980, ch. 2). The most distinctive feature of the scientific method is its explicit aim of generating testable predictions, so perhaps it is no surprise that our theoreticians keep getting better at devising empirically successful theories. Truth need not come into this picture and in fact might impede rather than support progress. Truth is likely to be complex and messy. Simple or 'idealized' theories make the essential job of creating testable predictions or intelligible explanatory accounts much easier.

Even if it is no miracle that science is so empirically successful, the second line of argument claims that this success should nonetheless lead to belief in the truth of our best theories. This argument depends upon both the correctness of IBE as a mode of reasoning and the appropriateness of its application in particular cases. The general idea is unexceptionable. When faced with some puzzling phenomenon we often accept the hypothesis which best explains it. If all the newspapers agree that the Maple Leafs lost yet again, then I ought to believe in their continued ineptitude. Why? Is it not possible that there is some conspiracy against the Leafs, or that an error was inadvertently repeated from newspaper to newspaper? While possible, these are bad explanations compared to the straightforward, and plausible, one.

Similarly, the defender of SR urges that the truth of our theories is the best explanation for their empirical success. To clinch the deal, it must be shown both that SR is indeed the best explanation and that this 'best' is sufficient to enjoin belief.

Obviously, the problem of underdetermination looms again when we turn to these tasks. If by 'best explanation' is meant only the best of what we have so far achieved then why should we embrace belief, given the existence of myriads of empirically equivalent, albeit unknown, contenders?

Worse still, there is a fundamental problem with IBE: it illicitly transforms explanatory into epistemic virtue despite the clear difference between them (see van Fraassen 1989, ch. 6).[2] For example, a prime explanatory virtue is simplicity. Yet there seems not the slightest reason to believe that simplicity tracks truth, and though one can debate the metric of simplicity, theories have clearly grown more complex over the last 400 years.

What are theories?

There is a hidden assumption, or presupposition, lurking in the anti-realist arguments. This is the unquestioned idea that the gauge of realism is the success of scientific theorizing.

What is especially interesting in the work of Nancy Cartwright and Ian Hacking[3] is the development of arguments for realism which not only avoid this presupposition but actively deny that realism should be theory based. To assess this we need to look briefly at how philosophers have understood theories.

Once again, the legacy of logical positivism looms large. A certain picture of theories dominated philosophy of science for the first two-thirds of the twentieth century. This picture was based upon early developments in mathematical logic, stemming from work of Gottlob Frege followed by that of Bertrand Russell and Alfred North Whitehead. By bringing to bear the purity of the logical notation and the power of first-order logic, it undeniably enabled philosophers to articulate and engage with a host of problems in the philosophy of science in far greater depth and with much greater clarity than ever before (see Suppe 1977).

However, the abstraction and simplification this 'logification' engendered tended to overemphasize features conducive to anti-realism. For example, once a theory is codified in pure logic it is easy to imagine dividing all the terms into two sets with specifiable relations such that one set becomes superfluous relative to the other. If we interpret these sets as 'observation terms' and 'theoretical terms', the result is Carl Hempel's 'theoretician's dilemma' (Hempel 1958), which purports to show that theoretical terms are superfluous or, at best, a mere convenience since their functions can be replaced with laws couched entirely in observational terms.

The casting of theories in logical form also distorted philosophical views of explanation. The famous deductive-nomological account reduced explanation to logical derivation. Many of its failings can be attributed to its acceptance of deductive consequence as the essence of the explanatory relation (see Suppe 1977 for extensive criticism).

This approach to theories, because of its assimilation of scientific theories to sets of first-order formulas, is sometimes called the 'syntactic view'. An alternative, labelled the 'semantic view' of theories and based upon early work of E. W. Beth and Patrick Suppes, has grown up in opposition. Broadly speaking, the semantic view replaces the formalist ideal with the claim that theories are specifications of the *models* to which the theory applies. One can freely use ordinary language and mathematics to delineate the relevant models.

The term 'model' is highly ambiguous, with three core meanings relevant to our discussion. Models could be set theoretic constructions which enable the definition of semantic rules and truth conditions. A different kind of model involves the specification of the 'state space' of the theory at issue. A familiar example is the phase space of classical mechanics. A third interpretation takes models to be possible physical systems, usually idealized in various respects but often such as can actually be built in at least approximate form or discovered in nature.

According to the semantic view, the truth of a theory depends on the relation between the models specified by the theory and reality. The simplest such relation is isomorphism. If there is a one-to-one map from one of a theory's models to reality, then the theory is true. Obviously, a demand for isomorphism is rather too stringent but it does allow for straightforward accounts of scientific progress. For example, consider the problem faced by eighteenth-century astronomers in accounting for the orbit of the moon (North 1995, 387ff.). The efforts of great mathematicians such as Euler, Laplace and Lagrange as well as the most able observers can be summarized as an attempt to construct a Newtonian gravitational model isomorphic to the actual motions of the moon. Although success was eventually achieved, the task was so difficult that amendments to Newton's theory were seriously entertained by both Euler and Laplace.

Theory realism

Theory realists maintain that it is the aim of science to produce true theories and that modern science has in fact produced theories which are at least close to the truth. From this point of view, there is no difference between SR and theory realism.

The original arguments against SR were aimed at the syntactic view of theories, and one might hope that adoption of the semantic view would undercut them. However, although the semantic view more properly represents the structure of theories and provides a better picture of how scientists devise and test them, the prospects for SR are not appreciably improved by its adoption. All the classic anti-realist arguments can be recast within the semantic view.

First, while the semantic view does not divide the regimented vocabulary of a theory into observational versus theoretical terms, one can reformulate the theoretician's dilemma. Our senses are limited and so our theories themselves predict a division of the world into the observable and unobservable. The models of the theory can then be internally articulated into observable and unobservable substructures. A theory is said to be 'empirically adequate' if its observable substructure is isomorphic to observable reality. The anti-realist argument is then apparent. The claim that a theory is empirically adequate is more probable than the claim that it is true.

It does not matter whether you take a semantic or syntactic view of theories; the pessimistic induction remains equally threatening.

The underdetermination argument clearly applies no less to the semantic view than the syntactic. It seems clearly possible that many distinct theories could share their empirical substructures and so be experimentally indistinguishable. To the anti-realist this stands as a virtue: the more theories which can grapple with the problem of explicating and predicting observable phenomena, the better. We can just pick and choose between them according to pragmatic concerns.

Thus the semantic view of theories does not by itself alter the dialectical situation. Anti-realists have a powerful set of arguments which tell equally against the traditional philosophy of science and the semantic view. Nonetheless, the emphasis on models can cast a new light on the debate if we bear in mind the various meanings of the term 'model'.

Entity realism

The year 1983 saw the publication of two books which dramatically altered the landscape of philosophy of science and presented a fundamentally new way to look at both theories and their appeal to models. These books were Nancy Cartwright's *How the Laws of Physics Lie* and Ian Hacking's *Representing and Intervening*. The core idea in these works is that theories are not 'where the action is' in the debate about scientific realism. Instead, Cartwright and Hacking suggested that realism should focus on *entities* rather than theories.

The *name* 'entity realism' entered philosophical discourse in 1979 when Brian Ellis described the view that things such as 'atoms and electrons are physical entities constitutive of the physical world. Science makes models of them and formulates laws governing the behaviour of these idealized particles in various idealized spaces. . . . [T]hese laws are not to be thought of as true generalized descriptions of the behaviour of the particles of physical reality, but statements about how these particles would behave in various kinds of idealized systems if they were exactly like the idealized particles of the models' (Ellis 1979, 45).

Although not the source of her reflections, Cartwright's work can be regarded as a detailed and comprehensive clarification and development of Ellis's suggestive remarks. The distinctive features of her account are, first, the defence of realism with respect to the entities discovered in the scientific endeavour and, second, the denial that scientific theories, and especially the laws of nature which they codify, are true. Crudely stated, the view is that while theoretical entities exist, theory realism is false. Hacking's views on entity realism (ER) are less worked out. If Cartwright represents the main column of the army, Hacking is more of a scout or sniper, ready to aid the attack but reluctant to fully join the ranks.

According to ER there are abundant and sound reasons for accepting the existence of the unobservable entities which scientists postulate. But these reasons are to a surprising degree independent of reasons for accepting the theories themselves. Part of the reason for this is that, according to ER, the nature of scientific theories has been deeply misunderstood by philosophers and scientists. Theories are not to be viewed as 'metaphysics with numbers' – they are not attempts to provide an accurate description of the natural world. Rather, they are a system of constraints on the construction of models.

ER thus takes a radical stance. In terms of the orthodox philosophical approaches outlined above, ER is indifferent to the possibility of axiomatizing theories in first-order language. Whether or not this is possible in particular cases has no bearing on the status of theories as representations of the world. Theories simply are not in the game of mirroring the world. Contrary to the semantic account, ER disputes the idea that the models of a theory are supposed to be isomorphic to reality. More important, the sense of the term 'model' appropriate for ER is the third noted above. Models are to be understood not as set-theoretic structures or abstract-state spaces but as more or less idealized physical systems.

A natural question is then, what good are theories? To understand ER's answer we need to distinguish between what Cartwright calls 'fundamental' and 'phenomenological' laws. These terms 'separate laws which are fundamental and explanatory from those that merely describe' (Cartwright 1983, 2). A simple example of a phenomenological law is the formula for hydrogen spectrum discovered by Jacob Balmer in 1884. Famously, Bohr was able to derive a generalized version of this 'excellent phenomenology' (Pais 1986, 164) from his early model of the atom. As Cartwright is at pains to point out, the phenomenological/fundamental distinction is not the observable/unobservable distinction. There are many phenomenological laws about unobservable entities. What distinguishes these laws is their attempt to describe measurable relationships.

Fundamental laws are the ones theory-realists think are getting at the deep structure of reality. The work required to generate a serviceable phenomenological law is regarded by them as merely the signature of complexity. Nature is a jumble of forces and constraints. But the ultimate truth shines through the mess in our fundamental laws. The entity realist sees this as backwards. The phenomenological laws are the true ones: they have confirmed instances, and they can be used to make predictions and build devices that work. The fundamental laws describe idealized systems. They are simple, they are useful, but they are false.

Cartwright's illuminating discussion of why airplanes fly perhaps makes this clearer (see Cartwright 1999, 27ff.). Don't fundamental laws of fluid mechanics govern lift and drag? Maybe, but if so, it is because we have produced a device which is close to the idealized models which fundamental laws describe: 'we build it to fit the models we know work' (Cartwright 1999, 28). We use theories as guides to model building. These models are ideal and not intended as true descriptions of

nature. However, there may well be systems in nature that come sufficiently close to the ideal model to enable description, explanation or
prediction. Our own solar system is close enough to an ideal Newtonian
model for many purposes.

This explains Cartwright's cryptic remarks about her famous example (originally Otto Neurath's) of the $1000 bill blowing about in St.
Stephen's Square. Obviously, the bill does not act much like the model
of a falling object, unlike a thrown brick or a rifle bullet, and the obvious reason is the interaction between the bill and the atmosphere.
Nonetheless, Cartwright appears to deny that the wind will have any
effect on the bill. She writes '[m]any will continue to feel that the wind
. . . must produce a force' (Cartwright 1999, 28). What can she mean?
That the wind does produce a force is a fact, as experience abundantly
attests. I think what she is getting at is simply the denial that the wind's
force can be scientifically modelled. There exists no fundamental law
for this situation, and of course, the complexities involved preclude the
possibility of devising a phenomenological law.

Here we must note a change in emphasis in Cartwright's views
between *How the Laws of Physics Lie* and *The Dappled World* (1999).
Compared to the earlier book, Cartwright is less concerned to show
that the scientific laws are 'lies' and more concerned to deny what she
calls 'fundamentalism'. This latter doctrine is the claim that there is a
domain of laws which govern all phenomena, in principle, and to which
all scientific laws must conform.[4] In denying fundamentalism, Cartwright
need not deny that there are true, or almost true, laws but only that
their application is severely limited and there is no essential hierarchy
of significance into which all laws must fall. Instead, she espouses 'the
view that the world in which we live – or rather the world given to
us through scientific investigation – is probably at best described by a
patchwork of laws with domains of limited range' (Steed et al., forthcoming). Significantly, this quote opens room for domains which are
perhaps not amenable to scientific investigation.

A particularly extreme form of a dappled world is bruited by Ian
Hacking in his Argentine fantasy:

> . . . God did not write a Book of Nature of the sort that the old Europeans
> imagined. He wrote a Borgesian library, each book of which is as brief as pos-
> sible, yet each book is inconsistent with every other. . . . For every book, there is

some humanly accessible bit of Nature such that that book, and no other, makes possible the comprehension, prediction and influencing of what is going on. (Hacking 1983, 219)

Outdoing Leibniz, Hacking continues: 'the best way to maximize phenomena and have the simplest laws is to have the laws inconsistent with each other, each applying to this or that but none applying to all'. If all of these laws are inconsistent with each other, then at most one could be true. It is more reasonable to regard them as all false but applicable to various aspects or regions of the world. I expect Hacking's light-hearted fantasy is not to be taken too seriously. It would be tempting to reply that each law should be formulated in conditional form with the antecedent describing its zone of applicability. Then all the laws would be universally true and mutually consistent. Such a reply would, however, fall into the trap of appealing to what are at best 'in principle' possibilities. The conditions of applicability would be so complex and convoluted as to be unspecifiable. This is the dilemma in which Cartwright and Hacking wish to place the theory realist. Either the fundamental laws are simple, elegant, actually expressible but false, or they are merely notional laws which are forever inaccessible to us. Science as we know it deals with, and can only deal with, the former sort of law.

This picture of overlapping, partially applicable laws reinforces another aspect of Cartwright's views. She maintains that the characterization of any aspect of nature, especially scientific experiments, requires application of diverse laws from separate and perhaps incompatible theories: 'neither quantum nor classical theories are sufficient on their own for providing accurate descriptions of the phenomena in their domain. Some situations require quantum descriptions, some classical and some a mix of both.' (Cartwright 2005, 194).

The metaphysical viewpoint attendant upon this view is potentially quite radical. Standard physicalism requires that the world have a fundamental structure which is, in principle, describable in terms of our best theories. If our 'best theories' form an inchoate and inconsistent set, more or less applicable across irreducibly diverse domains, then physicalism must be radically mistaken. This aspect of Cartwright's views have not gone unnoticed by scientists. In an extremely negative review of *The Dappled World*, Philip Anderson complains that Cartwright 'just does not get it' (Anderson 2001, 487), charges that her work 'will be useful to creationists' (487), calls her a 'social constructionist' (492) and claims that her presentation of specific examples, such as the BCS model of superconductivity, 'caricature[s] the real process of discovery and validation' (490).[5] Railing against being included as an ally in Cartwright's project in virtue of his famous musings on reductionism and solid state physics (Anderson 1972), Anderson quotes himself defending what he evidently regards as the only sensible metaphysics: '[t]he workings of our minds and bodies, and of all matter [. . .], are assumed to be controlled by the same set of fundamental laws, which [. . .] we know pretty well' (489). While the idea that Cartwright is a social constructionist is absurd, Anderson is right to see the threat to his idea that science presents the world as a seamless web.

Why believe in the entities?

So far we have considered ER's attack on theory realism. But in so far as this attack is successful, does it not also undercut our grounds for believing in the entities postulated by scientists? The concepts of these entities arise during theory construction, and without a theoretical superstructure they are literally unthinkable. Both Cartwright and Hacking have distinctive answers to this question.

Recall the IBE argument for SR: the best explanation for the success of our theories is their approximate truth. We also sketched the responses to the IBE argument. The claim that truth is the best explanation for the success of science is undermined by alternative accounts of science's empirical achievements. And even if the appeal to truth offered the best explanation of scientific success, the pessimistic induction suggests it nonetheless is not good enough to warrant belief.

Cartwright is alive to these criticisms and does not think highly of IBE. She accepts the distinction between explanatory and epistemic virtues, illustrating them with this analogy: 'I ask you to tell me an interesting story, and you do so. I may add that the story should be true. But if I do so, that is a new, additional requirement . . .' (Cartwright 1983, 91). Theoretical explanations are like interesting stories – perhaps they just *are* interesting stories, albeit frequently possessed of distinctive practical value.

Nonetheless, Cartwright thinks a version of the IBE argument can be deployed to license belief in theoretical entities. The core insight is that entities figure in causal processes. Cartwright's argument from 'inference to the most probable cause' (IMPC) works by first hypothesizing that a phenomenon is brought about via some entity's powers or 'causal capacities', then noting that this causal relationship stands as an explanation of the phenomenon's existence or its possession of relevant features. As with standard IBE, this explanation must be sufficiently persuasive or of high enough 'quality' to warrant belief in the postulated entities (see Cartwright 1983, 90ff.; for criticism, see Psillos 2008).

The crucial difference between IBE and IMPC, according to Cartwright, is that the latter is factive. It is impossible for X to cause Y unless X exists whereas S can explain Y even if S is not true. For example, Newton's theory of gravitation provides an explanation of the tides despite its falsity. Even if God appeared and entertained you with the interesting story 'that Schroedinger's equation provides a completely satisfactory derivation of the phenomenological law of radioactive decay', you 'still have no reason to believe in Schroedinger's equation' (Cartwright 1983, 93).

This asymmetry between IBE and IMPC can be questioned. 'X caused Y' is undeniably factive, but it does not immediately follow that 'X causally *explains* Y' is also factive. Perhaps to causally explain something is simply to posit an entity or property which, *if* it existed with such-and-such causal capacities, would cause the target phenomena.[6] It has been argued as well that although the asymmetry is real, it is merely conventional. On this account, it is just a parochial fact about the way we use language that makes cause-talk factive but explanation-talk non-factive (see Pierson and Reiner 2008). If so, no heavy duty philosophical conclusions can be drawn about realism.

Another criticism lies in the claim that entities vouchsafed to us by IMPC are conceptually dependent on theory. Another example of Cartwright's is that of the cloud chamber, in which charged particles leave a trail of condensation in a supersaturated medium. Cartwright says 'if there are no electrons in the cloud chamber, I do not know why the tracks are there' (Cartwright 1983, 99). But even if we grant that the cloud chamber gives reason to believe that something caused the track, how could we know that it is an electron? What an electron is seems dependent on theory. When J. J. Thomson discovered it, he may have envisaged a very small but otherwise fairly ordinary particle like a tiny marble – he called them corpuscles. The modern quantum mechanical view of the electron is radically different.

However, if the causal powers of theoretical entities involve the capacity to affect observers, albeit indirectly via instrumentation, we can, perhaps, call on the philosophy of language and in particular the causal theory of reference.[7] In oversimplified terms, this theory holds that the referent of a term, T, is the entity causally responsible for the production of instances of T. So if electrons, whatever they might be, were and are actually (part of) the cause of our use of the term 'electron', then everybody using that term is referring to the same thing, no matter how disparate the beliefs about electrons.

This philosophical escape hatch has the typical weakness of pure theory: it does not and cannot tell us anything about electrons. As Cartwright recognizes, the appeal to causation to license belief in the entities posited by science remains fundamentally beholden to theory. It is only in the context of theory that it is possible to form any judgment about the quality of a proffered causal explanation of some phenomena. This problem is endemic to thought, not merely science. Recall how readily causal explanatory hypotheses about the deeds of witches were once accepted. Odious but very widely held background 'theories' about the supernatural and its influence were crucial in allowing and sustaining belief in these entities and their putative causal powers. If we don't think the theories are more or less correct or if we haven't even heard of the theory or it hasn't been created yet, then we are going to be disinclined to accept, or even be incapable of formulating propositions about, the entities involved in the theory.

Here Cartwright borrows from the work of Hacking in the attempt to blunt this objection.[8] It is undeniable that theoretical knowledge is a prerequisite for making an inference to the most probable cause but Hacking argues there can nonetheless be relatively independent grounds available for the inference. This is possible if we can *manipulate* the entities in question. One way to think about the advantage manipulation provides is in terms of the causal theory of reference. Via manipulation we bring ourselves into causal contact with the entity in a way which is not theoretically mediated, no matter what theoretically informed cogitations led us to attempt the manipulation in the first place.

In fact, the argument is not very clear. One strand is transcendental: the possibility of coherent experimental practice requires realism about entities; experimental practice is (manifestly) coherent; therefore, ER (Hacking 1982, 73). A second strand more directly engages with manipulation. Hacking argues that sophisticated experimentation involves the construction of devices whose design, retaining the electron example, depends on 'a modest number of home truths about electrons' and whose purpose is to 'produce some other phenomenon that we wish to investigate' (Hacking 1982, 77).

The transcendental argument seems open to the 'fool's paradise' criticism (essentially, the underdetermination argument). That scientists have a certain world view without which their experimental practices make little or no sense to them does not provide very strong support for realism, for alternative world views with different entities might be equally coherent. Georg Stahl, Joseph Priestly and many others experimented extensively with what was called de-phlogisticated air (oxygen), phlogisticated air (a mixture of nitrogen and carbon dioxide) and the like and happily went about building experiments whose design and explication made sense – to them – only from within the phlogiston theory of combustion. Stahl offered a grand picture of nature in which the phlogiston cycle unified the three kingdoms of mineral, plant and animal (see Brock 1993, 78ff.). Use of phlogiston allowed chemists to perform all sorts of marvellous chemical transformations. For a while, it must have seemed that they were indeed manipulating phlogiston.

Hacking's argument depends on the fact that scientists are immersed in a picture of the world. The picture stems from the historical growth of scientific theory and the history of theoretical explanations of experimentation as these have been internalized by generations of scientists. When Thomas Kuhn famously claimed that pre- and post-revolution scientists lived in different worlds (with different entities), he was talking about this kind of immersion. The 'worlds' are 'the intentional correlate[s] of the conceptual framework through which I perceive and conceive the world' (van Fraassen 1980, 81). Coherence of experimental practice is a requirement for the retention of the conceptual framework itself.

On the other hand, I think the manipulation strand of the argument can be regarded as independent of the transcendental one. While theoretical immersion is required for experimentation, the evidence for certain entities does not require one to believe in those theories (hence, Hacking's remark about 'home truths'). In fact, the ability to manipulate the electron in various ways is used to test hypothetical regions of theory. It is worth remembering that the basic proposition that there are unobservable entities in nature is at most minimally theoretically loaded – there are only philosophical objections to it – and is, outside philosophy lecture rooms, universally accepted. This is the fundamental disconnect. Scientists have established procedures for investigating realms of being absolutely invisible to human senses. They possess no doubts about the general feasibility of this procedure which has long borne fruit of both theoretical and practical importance. Untutored thinkers intuitively agree with the cogency of this procedure.

The distinction between what might be called 'theory free' versus 'theory laden' entities is one of degree, with super-hypothetical entities like the Higgs boson at one end and, say, the adenovirus or electrons at the other. Hacking's point is that it is the system of intervention and manipulation which grades this continuum. As this system becomes more extensive and more fully linked to other practices and technology, the existence and causal powers of the entities in question become increasingly hard to disavow, independent of allegiance to theory. For example, we now use genetically modified adenoviruses to implant novel genes in people for therapeutic purposes, and many patents exist, dating back decades, for sources of spin-polarized electrons.

If Hacking is correct, Cartwright can answer the criticism that there is no way to assess existence claims apart from theory acceptance.

Problems for entity realism

Many difficulties could be raised against ER. A frequent objection, already considered, is that there is no viable way to disentangle commitment to theory from grounds for belief in the existence of entities which are unthinkable without theory (see Morrison 1990). Another possible objection discussed above claims that IMPC fares no better than basic IBE, so all of Cartwright's and Hacking's criticisms of IBE-based realism can be transferred to the case of ER with equally damaging effect.

I want to conclude by considering a couple of different objections that focus on Hacking's manipulability condition and Cartwright's appeal to the causal capacities of entities.

According to Cartwright, the traditional view that laws explain the causal proclivities of nature's entities has things backwards. The truth is that the causal capacities of entities, working together in complex ways, generate natural structures and processes which are amenable to a description couched in terms of natural laws and occasionally conspire together in ways that sustain practical prediction and stable technology (see Cartwright 1989). These conspiracies of nature or artifice Cartwright calls 'nomological machines': 'a fixed (enough) arrangement of components, or factors, with stable (enough) capacities that in the right sort of stable (enough) environment will, with repeated operation, give rise to the kind of regular behaviour that we represent in our scientific laws' (Cartwright 1999, 50).

One question this view raises is whether there is any real prospect of retrieving all the fundamental laws from entities and their capacities or whether at least some scientific laws have to be regarded as in some way self-standing. Many principles of basic science do not seem capacity based. Consider the Pauli exclusion principle, which states that no two fermions can be in the same quantum state. Without this law of nature, quantum mechanics would lose most of its explanatory power, especially in chemistry. Unfortunately, there is no hint of anything which forces the truth of the exclusion principle in the causal capacities of the fermions. It would thus have to be added as a brute, primitive capacity, but a very peculiar one that is intrinsically relational and 'non-local'. An interesting possible route towards a causal account of the exclusion principle can be found in modern versions of Bohmian quantum mechanics (see Riggs 2009, ch. 6). However, this account requires appeal to the quantum potential – a non-local holistic field which encompasses the entire universe – and this is not an acceptable entity in Cartwright's terms. Certainly it is not something we can manipulate. Bohm's approach also entails a radical interpretation of quantum mechanics which seems entirely at odds with her vision of a 'dappled world'.

It would not be difficult to find other examples of principles which do not seem to be capacity based.

A more general problem for ER is that the capacities approach might support a return to a kind of physics fundamentalism. This follows from the idea that there is a set of entities which are in themselves individual nomological machines and are such that the capacities of all other entities supervene on the capacities of this distinguished set. The fundamental entities of physics can perhaps be viewed as both existent and exhaustively described by the models of physical theory. That is, the electron is a real thing, but it is perfectly and fully characterized by physical theory. ER accepts the existence of the entities of fundamental physics (some frontier entities remain hypothetical). One might then argue that the structure of physical theory suggests that the causal capacities of the set of elementary entities determines all other causal powers. The following remarks of Hacking's, on the standard model of physics, express my worry. This model is 'the fulfillment of Newton's program, in the *Optics*, which taught that all nature would be understood by the interaction of particles with various forces that were effective in attraction or repulsion over various distances' (Hacking 1982, 78).

On this view, which seems at odds with Cartwright's overall vision, although no theory provides a set of true fundamental laws in which all physical systems can be modelled, it is nonetheless true that all of nature is completely determined by the causal capacities, and particular activities, of a small number of elementary entities. These entities are so simple and circumscribed in their properties that they are fully described by the theories which posit them. Theory realism thus returns through the back door as the true depiction of the basic entities which determine everything else in nature.

There seem to be only two possible responses to this worry. One is to deny that the fundamental entities posited by basic physics deserve our ontological endorsement, but this seems quite at odds with the general thrust of ER. After all, a good many of these elementary entities meet all the conditions which Cartwright and Hacking demand for belief. The second possible response is even bolder. It is to deny that the theories which describe the fundamental entities are fully correct. Instead, it regards *every* facet of the universe as embodying complexity within complexity and the descriptions provided by science of the entities we call fundamental are, in line with other areas of science, mere idealized versions of what exists in nature. On this rather Leibnizian view, the apparent indistinguishability of the elementary entities is a mere artefact of experimental efforts designed to transform some microscopic portion of the world into one of Cartwright's nomological machines.

Turning to Hacking, the manipulability criterion for belief in the reality of theoretical entities leads to some obvious objections. One is the distressing habit of non-existents to meet the criterion. Priestley manipulated – in his own mind – phlogiston. The point is not the obvious one that we can be wrong in our hypotheses about entities no less than theories, but that the 'intentional correlate' of a false theory can provide a world view that sustains practice (see Seager 1995). The 'home truths' or 'low-level causal properties' (Hacking 1983, 174) of real entities to which Hacking appeals are thus just congealed theory – they reflect theoretical propositions so vital to ongoing theorizing and experimentation that there is no real prospect of their disappearing except by truly radical changes in fundamental science. Thus ER and theory realism may not be so different as Cartwright and Hacking would like them to appear.

There is also a worrying analogy between ER and scientific antirealism. Recall that for the typical anti-realist the notion of observability plays a crucial role by dividing the world into zones worthy or unworthy of belief. Realists decry the appeal to such a parochial and arbitrary distinction. The concept of manipulability is formally analogous to that of observability, however. There is a clear prior intuition that there are very likely to be unmanipulable entities in the world, just as it seems intuitively compelling that there are unobservables. As with observability, it is our scientific knowledge that tells us where the fuzzy line is between manipulable and unmanipulable. And this seems a rather limited or local fact which essentially depends on our own nature. Just as some things are too small for us to observe, some things are just too far away for us to manipulate. Other things interact so weakly with the things we can manipulate that it seems difficult to maintain that we can manipulate them.

An interesting case of this latter category is the neutrino, a particle which interacts with matter only via the weak force. Untold numbers of neutrinos constantly stream through the earth and us with essentially no effect. Yet Hacking allows neutrinos into the circle of manipulability because we can, under arduously contrived experimental circumstances, detect a tiny portion of the neutrino flux and calculate certain other physical properties from these measurements. For example, in 1987 the coincidence of a new supernova and the detection of 24 neutrinos in three separate neutrino 'observatories' could be used to confirm some aspects of theoretical models of supernovae. This entirely passive process evidently counts as 'intervention', because 'we are plainly using neutrinos to investigate something else' (Hacking 1982, 71).

If this objection is cogent we should expect to find Hacking struggling to be an *anti*-realist about certain scientifically certified entities that fall outside the class of the manipulable. And we do. Hacking has expressed reservations about black holes on the grounds that they are both unobservable and unmanipulable. He admits to '. . . a certain scepticism, about, say, black holes. I suspect there might be another representation of the universe, equally consistent with phenomena, in which black holes are precluded'. In support of the unobservability of black holes, Hacking goes on to write that a black hole is 'in principle unobservable' and '[a]t best we can interpret various phenomena as being due to the existence of black holes' (Hacking 1983, 275). These remarks are somewhat curious. About observability, the distinction between observation 'directly' and only via some effect is murky at best. In fact, it threatens to break down completely once we allow use of exotic detection devices to count as observation (such as the use of arrays of photomultiplier tubes immersed in dry cleaning fluid to 'see' neutrinos).

Hacking's denigration of black holes on the basis of there being a possible representation (theory?) without them seems to be in tension with the spirit of ER. Does a black hole have to harbour a singularity at its heart? If so, then it seems likely that black holes are flat out impossible and are mere mathematical artefacts. But ER allows for the existence of entities which fail to be completely or fully described by the theory which engenders their concept – that is one of the main points of the view. Are there regions of nature where the local gravitational field is strong enough to make the escape velocity greater than the speed of light? We have abundant evidence there are. For example, extended observations of the core of our galaxy have provided infrared images of a star in orbit around our central supermassive black hole which virtually eliminate alternative possibilities (see Schodel et al. 2002).

But of course we cannot manipulate or interfere with black holes. Or can we? It is now recognized that, if certain speculative but wellregarded theories which entail that there are extra-spatial dimensions are correct, the Large Hadron Collider may generate microscopic black holes in copious quantities. The decay of such black holes, by emission of Hawking radiation, will be unambiguous, and the decay products will be useful for testing 'the higher dimensional Hawking evaporation law' and will serve to 'determine the number of large new dimensions and the scale of quantum gravity' (Dimopoulos and Landsberg 2001). If experimental access to micro black holes comes about, will the ontological lesson carry over to astrophysical black holes?

On the other hand, micro black holes are extremely speculative, and it is more likely that black holes will remain solely in the domain of astrophysics and astronomy. We might still 'observe' them via the gravitational waves they will emit under certain theoretically specifiable conditions. The requisite observatory – the Laser Interferometer Gravitational-Wave Observatory (LIGO) – is already in place, though not yet functioning at its full potential and with as yet no detection of gravity waves. Is the difference between detecting the extreme orbital dynamics of a star which is very near a supermassive black hole and detecting gravitational waves from an orbiting pair of such objects so different that one counts towards ontology and one does not? Once we are routinely observing gravitational waves, LIGO will become just another kind of telescope, but its operation and the interpretation of its findings will not make much sense without the background belief in the sources of the radiation we are exploiting.

This line of criticism can be extended (Shapere 1993). I think it would be natural for Hacking to reply that he is tracing the admission of entities into the realist's circle of acceptance and that interference, manipulation and the 'use' of the entity for investigation into other parts of nature are the harbingers of the acceptance of an entity as real. This process is generally complex, convoluted and lengthy.

While this is true, the emphasis on manipulation and interaction seems untoward and overdone. The consolidation of scientific ontology is more liberal and promiscuous than Hacking and Cartwright allow. Entities lose their hypothetical status as a scientific world view solidifies. One source of this solidification is a rich history of experimental practice and a theoretical interpretation of it. Another is the integration of this practice into the picture of the world which our fundamental theories suggest to working scientists who possess some philosophical imagination (perhaps only a small fraction of working scientists in general). Thus it is unclear that entity realism can be cleanly distinguished from theory realism. Rather, ontology is fixed by the interaction of theory, experimentation and philosophical reflection.

William Seager
University of Toronto

Notes

1 So, while it is undeniable that '. . . an inductive skeptic could offer a pessimistic induction against anti-realists and realists alike' since strictly speaking most 'theories eventually turned out not to be empirically adequate' (Lange 2002, 282), the anti-realist can hold that the older theories are as empirically adequate as they ever were. Such claims have substantial content. On the other hand, one cannot really make sense of the idea that the old theories were 'partially true' unless one illegitimately reinterprets truth in terms of degree of empirical success.

2 An interesting assessment of van Fraasen's viewpoint plus an argument for the compatibility of IBE and Bayesianism can be found in Weisberg (2009).

3 It is obviously impossible to discuss all of Hacking's manifold contributions in the space of this article, which is rather devoted to the confluence of Cartwright's and Hacking's work on scientific realism and the role of theories therein. Here I can only note Hacking's work in the history of philosophy (Hacking 1973), the philosophy of language (Hacking 1975b), the history and philosophy of probability and statistics (Hacking 1975a; Hacking 1990), the history of mental illness and philosophical psychology (Hacking 1995, 1998) and social constructionism in science studies (Hacking 1999).

4 For a vigorous defence of physics fundamentalism against Cartwright, see Hoefer (2008).

5 For Cartwright's reply to Anderson's 'grossly inaccurate' review, see Cartwright (2001).

6 One might also question whether explanation in general is really ontologically innocent; see Seager (1987).

7 The modern version of the causal theory of reference mostly stems from the work of Saul Kripke (1980) and Hilary Putnam (see, e.g., 1975). For Hacking's view of its role in entity realism, see Hacking (1982).

8 See Hacking (1983), ch. 16, or Hacking (1982). For criticism, see Reiner and Pierson (1995) and Resnik (1994).

References

Anderson, Philip (1972). 'More is different', *Science*, 177(4047): 393–6.

— (2001), 'Science: A "dappled world" or a "seamless web"?', *Studies in History and Philosophy of Modern Physics*, 32(3): 487–94.

Brock, William (1993), *The Norton History of Chemistry*. New York: Norton.

Carrier, Martin (2004), 'Experimental success and the revelation of reality: The miracle argument for scientific realism', in M. Carrier, J. Roggenhofer, G. Küppers, P. Blanchard et al. (eds), *Knowledge and the World: Challenges Beyond the Science Wars*. Berlin: Springer.

Cartwright, Nancy (1983), *How the Laws of Physics Lie*. Oxford: Oxford University Press (Clarendon).

— (1989), *Nature's Capacities and Their Measurement*. Oxford: Oxford University Press.

— (1999), *The Dappled World*. Cambridge: Cambridge University Press.

— (2001), 'Reply to Anderson', *Studies in History and Philosophy of Modern Physics*, 32(3): 495–7.

— (2005), 'Another philosopher looks at quantum mechanics, or what quantum theory is not', in Y. Ben-Menahem (ed.), *Hilary Putnam, Contemporary Philosophy in Focus*. Cambridge: Cambridge University Press, pp. 188–202.

Dimopoulos, Savas and Greg Landsberg (2001), 'Black holes at the Large Hadron Collider', *Physical Review Letters*, 87(16): 161–2.

Duhem, Pierre (1914/1991), *The Aim and Structure of Physical Theory*, P. Wiener (trans.). Princeton, NJ: Princeton University Press.

Ellis, Brian (1979), *Rational Belief Systems*. Oxford: Blackwell.

Hacking, Ian (1973), *Leibniz and Descartes: Proof and Eternal Truths*. Oxford: Oxford University Press.

— (1975a), *The Emergence of Probability: A Philosophical Study of Early Ideas About Probability, Induction and Statistical Inference*. New York: Cambridge University Press.

— (1975b), *Why Does Language Matter to Philosophy?* Cambridge: Cambridge University Press.

— (1982), 'Experimentation and scientific realism', *Philosophical Topics*, 13(1): 71–87.

— (1983), *Representing and Intervening: Introductory Topics in the Philosophy of Natural Science*. Cambridge: Cambridge University Press.

— (1989), 'Extragalactic reality: The case of gravitational lensing'. *Philosophy of Science*, 56(4): 555–81.

— (1990), *The Taming of Chance*. Cambridge: Cambridge University Press.

— (1995), *Rewriting the Soul: Multiple Personality and the Sciences of Memory*. Princeton, NJ: Princeton University Press.

— (1998), *Mad Travelers: Reflections on the Reality of Transient Mental Illnesses*. Charlottesville: University Press of Virginia.

— (1999), *The Social Construction of What?* Cambridge, MA: Harvard University Press.

Hempel, Carl (1958), 'The theoretician's dilemma', in Herbert Feigl, Michael Scriven and Grover Maxwell (eds), *Minnesota Studies in the Philosophy of Science*, vol. 3. Minneapolis: University of Minnesota Press.

Hoefer, Carl (2008), 'For fundamentalism', in Luc Bovens, Carl Hoefer and Stephan Hartmann (eds), *Nancy Cartwright's Philosophy of Science*. London: Routledge, pp. 167–94.

Kripke, Saul (1980), *Naming and Necessity*. Cambridge: Cambridge University Press.

Cartwright and Hacking

Lange, Marc (2002), 'Baseball, pessimistic inductions and the turn¢ *Analysis*, 62: 281–5.

Laudan, Larry (1981), 'A confutation of convergent realism', *Philosophy of Science*, 48: 19–49.

Morrison, Margaret (1990), 'Theory, intervention and realism', *Synthese*, 82(1): 1–22.

North, John (1995), *The Norton History of Astronomy and Cosmology*. New York: Norton.

Pais, Abraham (1986), *Inward Bound*. New York: Oxford University Press.

Pierson, Robert and Richard Reiner (2008), 'Explanatory warrant for scientific realism', *Synthese*, 161: 271–82.

Psillos, Stathis (2008), 'Cartwright's realist toil: From entities to capacities'. In Luc Bovens, Carl Hoefer and Stephan Hartmann (eds), *Nancy Cartwright's Philosophy of Science*. London: Routledge.

Putnam, Hilary (1975), 'Explanation and reference'. In *Mind, Language and Reality: Philosophical Papers*, vol. 2. Cambridge: Cambridge University Press.

Reiner, Richard and Robert Pierson (1995), 'Hacking's experimental realism: An untenable middle ground', *Philosophy of Science*, 62: 60–9.

Resnik, David (1994), 'Hacking's experimental realism', *Canadian Journal of Philosophy*, 24: 395–412.

Riggs, Peter (2009), *Quantum Causality: Conceptual Isuses in the Causal Theory of Quantum Mechanics*. Dordrecht: Springer.

Schodel, R., et al. (2002), 'A star in a 15.2-year orbit around the supermassive black hole at the centre of the Milky Way', *Nature*, 419: 694–6.

Seager, William (1987), 'Credibility, confirmation and explanation', *British Journal for the Philosophy of Science*, 38(3): 301–17.

— (1995), 'Ground truth and virtual reality: Hacking vs. van Fraassen', *Philosophy of Science*, 62(3): 459–78.

Shapere, Dudley (1993), 'Astronomy and antirealism', *Philosophy of Science*, 60(1): 134–50.

Steed, Sheldon, Gabriele Contessa, et al. (forthcoming), 'Keeping track of Neurath's bill: Abstract concepts, stock models and the unity of classical physics', in O. Pombo (ed.), *The Unity of Science: Essays in Honour of Otto Neurath*. Berlin: Springer, http://personal.lse.ac.uk/cartwrig/PapersGeneral/Keeping track on Neurath bill.Cartwright Contessa Steed.pdf.

Suppe, Frederick (1977), 'The search for philosophic understanding of scientific theories', in Frederick Suppe (ed.), *The Structure of Scientific Theories*. Urbana: University of Illinois Press, pp. 3–241.

van Fraassen, Bas (1980), *The Scientific Image*. Oxford: Oxford University Press (Clarendon).

Weisberg, Jonathan (2009), 'Locating IBE in the Bayesian framework', *Synthese*, 167: 125–43.

CHAPTER 11

FEMINIST CRITIQUES: HARDING AND LONGINO
Janet Kourany

In a now infamous speech to a National Bureau of Economic Research Conference back in 2005, former Harvard University president (and former director of the National Economic Council) Lawrence Summers offered his hypothesis regarding the ongoing gender gap in science and engineering at top universities and research institutions (see Summers 2005). Women, Summers suggested, are still not flourishing in science the way men are because women just don't have what it takes. First, they are not motivated enough to do the 24/7 job that high scientific achievement demands. Second, they are not bright enough – not gifted enough in the analytical skills required to realize that high scientific achievement. And, oh yes, third, they also contend with, here and there, some slightly negative socialization and some small amount of residual discrimination. The upshot is that efforts to close the gender gap in science are ultimately doomed to failure, and hence, resources aimed at that closure had best be directed elsewhere. Interestingly, this speech was presented at a conference whose title was 'Diversifying the Science and Engineering Workforce'.

Summers's speech unleashed a firestorm of controversy. Some applauded Summers's refreshing candour, his willingness to say out loud what was doubtless on everyone's mind. Most, however, sharply criticized Summers for his lack of attention to what were then the latest results of research – the MIT study of 1999, for example, and the succession of studies thereafter that documented serious continuing discrimination against women scientists; the studies that linked women's comparative lack of achievement in science to women's socialization and the still current stereotypes and gender roles that hold women back; the studies that demonstrated women's equal potential in mathematics; the cross-cultural studies of women's success in science in other times and places; and so on (for accounts of some of these studies, see MIT Committee on Women Faculty in the School of Science 1999; Gallagher and Kaufman 2005; Wylie, Jakobsen and Fosado 2007; Ceci and Williams 2007, 2010).

The whole debate, however, was, and continues to be, strangely incomplete. Consider, for example, the questions habitually asked and not asked. It is always asked whether women are as smart or as gifted for scientific endeavours as men. It is never or rarely asked whether men are as smart or as gifted for scientific endeavours as women – whether men scientists, for example, would do as well as women scientists do now if the men had to deal with the socialization and discrimination and stereotype threat and other problems with which the women scientists regularly deal. Again, it is always asked whether women scientists who are mothers or who will be mothers or who might be mothers can be as single-minded and focused and detached from family and community concerns as men. It is never or rarely asked whether such ongoing stereotypically masculine detachment from family concerns is really required for high scientific achievement or whether it is ever actually justified. Most important, it is always asked whether women, should they be given equal opportunities in science with men, will be able to make the same kinds of contributions to science as men. Never considered is the possibility that women might make *different* contributions to science than men, contributions that might augment or correct or enlarge the contributions of the men. Indeed, it is insinuated that the science men have produced is fine as is, or can be made fine by the efforts of still other men. It is insinuated that the contributions of women are simply not needed. And this is shocking in view of the critiques of many areas of science provided by women scientists during the last thirty years, critiques that reveal quite striking failures and limitations in these sciences that women scientists have done much to rectify. In what follows, we shall take a quick tour of some of these critiques in order to make clear just what the situation has been. We shall then be in a position to appreciate the new understandings of science these critiques suggest – new understandings developed by feminist philosophers of science

Sandra Harding and Helen Longino – and the changed complexion the Summers debate then takes on.

Reframing the debate, part 1: factoring in what women scientists contribute

Start with medical research, especially in the United States. What have women scientists revealed about medical research? Simply that most people tended to be neglected in medical research – in both basic and clinical medical research – until the 1990s (for what follows, see Rosser 1994; Weisman and Cassard 1994; Gura 1995; Mann 1995; Meinert 1995; Sherman, Temple and Merkatz 1995; Schiebinger 1999). I will focus here on women in particular, though much can be said about the various groups of disenfranchised men as well. Regarding women, then, three of the more egregious areas of neglect in medical research were heart disease, AIDS and breast cancer – despite the fact that heart disease is the leading cause of death among women, AIDS is increasing more rapidly among women than among the members of any other group and breast cancer has for years been the most frequently occurring cancer in women. The result was that these diseases were often not detected in women – often not even suspected – and not properly managed when they were detected.

Consider just heart disease. Slightly more than one out of every two women in the United States will die from cardiovascular illnesses. Yet until the 1990s heart disease was defined as a male disease and studied primarily in white, middle-aged, middle-class men. The large, well-publicized, well-funded studies of the past are illustrative: the Physicians' Health Study, whose results were published in 1989, examined the effect of low-dose aspirin therapy on the risk of heart attack in 22,071 male physicians; the Multiple Risk Factor Intervention Trial (MR. FIT), whose results were published in 1990, examined the impact of losing weight, giving up smoking and lowering cholesterol levels on the risk of heart attack in 12,866 men; the Health Professionals Follow-Up Study, whose results were also published in 1990, examined the relation of coffee consumption and heart disease in 45,589 men. These studies were no exceptions: in a 1992 *Journal of the American*

Medical Association analysis of all clinical trials of medications used to treat acute heart attack published in English-language journals between 1960 and 1991, for example, it was found that fewer than 20 per cent of the subjects were women. 'When I began studying cardiovascular disease, it was all about women taking care of their husbands' hearts', recalls cardiologist Bernadine Healy, who, as the first woman director of the National Institutes of Health from 1991 to 1993, did much to change the situation. 'Heart disease in women was either trivialized or ignored for years' (quoted in Gura 1995, 771).

The consequences of that neglect were far from trivial, however. Since women were not researched along with the men, it was not discovered for years that women differed from men in symptoms, patterns of disease development and reactions to treatment. As a result, heart disease in women was often not detected, and it was often not even suspected. What's more, it was not properly managed when it was detected. Drug treatments were a particularly glaring example. Drugs that were beneficial to many men caused problems in many women. For example, some clot-dissolving drugs used to treat heart attacks in men caused bleeding problems in women, and some standard drugs used to treat high blood pressure tended to lower men's mortality from heart attack while they raised women's mortality. Furthermore, the dosage of drugs commonly prescribed for men was often not suitable for women. Some drugs (e.g., antidepressants) varied in their effects over the course of the menstrual cycle, while others (e.g., acetaminophen, an ingredient in many pain relievers) were eliminated in women at slower rates than in men. Studying only, or primarily, men resulted in failures to prescribe appropriate kinds and doses of drugs for women, as well as failures to offer other treatments (cardiac catheterization, coronary angioplasty, angiography, artery bypass surgery) at appropriate times. And it limited women's access to experimental therapies.

The effects of the old exclusions still linger. For example, women consume roughly 80 per cent of the pharmaceuticals used in the United States, but they are still frequently prescribed drugs and dosages devised for men's conditions and average weights and metabolisms. As a result, adverse reactions to drugs occur twice as often in women as in men. Indeed, 'the net effect', concludes historian of science Londa Schiebinger, 'is that women suffer unnecessarily and die. . . . Not only are drugs developed for men potentially dangerous for women; drugs potentially beneficial to women may [have been] eliminated in early testing because the test group [did] not include women' (Schiebinger 1999, 115).

Consider, for a second example, archaeology. What have women scientists revealed here? Archaeology is a field in which, traditionally, the search for origins and key developments in human evolution defines the 'big' questions, the prestigious questions. It is this search, in fact, that allows archaeologists to structure their discipline, determine career success and make their sometimes sensational statements about human nature and human society when presenting the results of their research. But until the 1990s, what archaeologists recognized as the 'hallmarks' of human evolution – tools, fire, hunting, food storage, language, agriculture, metallurgy – had all been associated with men. Agriculture, for example. Although women had always been firmly associated by archaeologists with plants – both with gathering them (before the emergence of agriculture) and with cultivating them (after) – when archaeologists turned to the profoundly culture-transforming shift in subsistence practice represented by the invention of agriculture, women disappeared from discussion. Until the 1990s, dominant explanations of the emergence of agriculture in the Eastern Woodlands of North America posited either male shamans and their ceremonial use of gourd rattles as the catalysts for this transition, or plants' 'automatic' processes of adaptation to the environmentally disturbed areas of human living sites (in which case the plants essentially domesticated themselves). According to these explanations, in short, either men invented agriculture, or no one did (Watson and Kennedy 1991). 'We have had, it seems, little problem in attributing a great deal of the archaeological record to men (the more salient stone tools, the hunting of big game, the making of "art", the development of power politics, the building of pyramids and mounds, the invention of writing by priests or temple accountants, domesticating gourds in order to have them available for shamans' rattles, etc.)' (Conkey 2008, 49). In addition, archaeologists have had little problem leaving out of the archaeological record what might easily, even stereotypically, have involved the experiences and contributions of women, such as midwifery and mothering practices, the socialization and gendering of children, sexual activities and relationships and the social negotiations surrounding death, burial and inheritance, topics that also hold enormous importance for the evolution of humans (for the beginnings of change on such topics, see Meskell 1998; Joyce 2000; Schmidt and Voss 2000; Wilkie 2003; Baxter 2005). As a result of this mode of representation of the past, this persistent association of men with the great turning points of human evolution, man as active, instrumental (as in man the toolmaker), man as provider, man as innovator, man as quintessentially human, were made to seem natural, inevitable. At the same time, woman as outside the domain of innovation and control, woman as not active (i.e., passive) and less than quintessentially human, were made to seem natural and inevitable as well, and thus capable of explaining (and justifying) the claims of women's inferiority and lesser potential we still find today (Conkey and Williams 1991; cf. Conkey 2008).

Consider, for a third and final example, what women scientists have revealed about economics (for what follows, see Waring 1992, 1997; Ferber and Nelson 1993, 2003; Nelson 1996a, b; Estin 2005). The central concept in current mainstream economics ('neoclassical' economics) is that of 'the market', a place where rational, autonomous, self-interested agents with stable preferences interact for the purposes of exchange. These agents may be individual persons or collectives of various kinds, such as corporations, labour unions and governments. The agents, in either case, exchange goods or services, with money facilitating the transactions; and the tool of choice for analyzing these transactions is mathematics. Indeed, high status is assigned in economics to formal mathematical models of rational choice. What tends to remain invisible, however, or inadequately treated, are women.

Take women's experiences in the family. Since the focus in mainstream economics is on the 'public' realm (industry and government), 'private' collectives, such as the family, tend to get scant attention. And since the prototype for economic agents is individual persons, and masculine persons at that,[1] when families are attended to, they are most commonly treated as if they were individuals themselves, with all their internal workings a 'black box'. Or they are treated as if they had a dominant 'head' who makes all the decisions in accordance with 'his' own (perhaps altruistic) preferences. Either way of treating the family leaves women invisible as agents in their own right in the family. More recently, however, families have been treated by some economists as (cooperative or non-cooperative) collective decision-making partnerships. But since, here as elsewhere in mainstream economics, the focus is on simplified mathematical models portraying the interactions of rational, autonomous agents, these collective decision-making partnerships end by being models of marital *couples*. Children, not yet fully rational, certainly not autonomous and threatening to the tractability of the models, are either conceptualized as 'consumption goods' or not conceptualized at all. Left invisible, therefore, are women in the family as caregivers, as agents who historically[2] have borne the bulk of the responsibility for the nurturance and education of children and the care of the sick and elderly. The upshot is that women's needs and priorities in families are left invisible and, with them, the impact on women and their charges of public policies:

> Any model of the effect of price changes, or taxes or transfers on family behavior must, implicitly or explicitly, rely on a theory of how families function. Beyond the social scientists' need to understand, lies the policy-makers' need to make wise policies. Better knowledge about what is happening in the family could improve policies related to child poverty and child support, household-sector savings rates, welfare and job training, the tax treatment of dependents and family-related expenses, social security, elder care, healthcare, and inheritance taxation, to name a few areas. (Nelson 1996, 60)

Mainstream economics obscures women's situation in families in other ways as well. Indeed, it essentially offers only one theory of why the household is organized in the way it is. According to this theory,

> individuals decide whether to marry, have children, or divorce based on a comparison of the direct and indirect benefits and costs of different actions. It conceives of people in family relations as rational, utility-maximizing actors, facing a range of choices with limited supplies of time, energy, wealth, and other resources. It imagines that for most goods there are substitutes, although recognizing that these may not be perfect, and postulates that individual actors have an ability to choose freely. (Estin 2005, 434–5)

Finally, this one theory provided by mainstream economics describes the gender-based division of labour in households as 'efficient' in so far as that division of labour allows a family to generate a maximum level of utility with a given set of resources – describes it as efficient even though that division of labour often results from factors such as discrimination against women in the workplace, often involves wives bearing much heavier loads of work and family responsibilities than their husbands and often leaves wives vulnerable to domestic abuse and other effects of lower status in marriage and financial problems with divorce. Wedded to efficiency as a normative ideal, however, this one theory provided by mainstream economics does not invite exploration of policies aimed at restructuring the traditional household and how such policies might impact other problem areas for women, such as the gender wage gap. As a result, mainstream economic theory, with its focus on economic efficiency, serves to further reinforce and justify a social order systematically harmful to women.

The above represents only a few of the areas of science critiqued by women scientists during the last three decades. Still, information relevant to the Larry Summers debate can be gleaned from these few examples. First, the people who have carried out the critiques are overwhelmingly women scientists active in the areas of science critiqued. That is to say, it is overwhelmingly women scientists who have diagnosed the gender- and sometimes also race- and class-related failures and limitations of these research areas. Second, it is overwhelmingly men's science that is the object of these critiques. And third, women scientists have not only diagnosed the gender-related failures and limitations of these research areas but have also done much work toward rectifying them.

For example, women archaeologists have challenged the old origins stories – for example, the 'man the toolmaker' story – as resting on assumptions about the division of labour between the sexes applicable only quite recently and only in European and American cultures. At the same time they have worked to include women in these stories as active participants – for example, as active developers of pottery, an invention of major historic significance, and as active domesticators whose agricultural feats provided the staples of ancient diets. Most important, women archaeologists have opened up new questions for research, such as how tools were used in activities other than big-game hunting (the traditional tool question in archaeology). They have asked how tools were used in food preparation and leatherworking and grain harvesting and woodworking; they have asked what early people usually ate and what the economic and cultural goals of toolmaking were. And they have asked other new questions as well, questions that explore men's activities as well as women's (e.g., see Conkey 2003, 2008).

For their part women medical researchers have not only worked to ensure equal attention to women's and men's needs in medical research but have also begun the difficult but necessary re-conceptualization of what such equality of attention requires. Indeed, medical researchers, such as NIH reproductive endocrinologist Florence Haseltine, have presided over a shift in women's health research from merely reproductive research (involving attention to childbirth, contraception, abortion, premenstrual syndrome, breast and uterine cancers and the like) to more general health research (involving attention to women's distinctive physiology), and this has been critical to improving health care for women. Other medical researchers – such as Adele Clarke, Elizabeth Fee, Vanessa Gamble and Nancy Krieger – have in turn moved the understanding of health research toward a broader social model that does more than focus on disease management and biochemical processes. This broader model takes into account how health and disease are produced by people's daily lives, access to medical care, economic standing and relations to their community. It thereby takes into account the differing health needs of women and men of different races and ethnic origins and classes and sexual orientations (see, e.g., Moss 1996 and Schiebinger 1999).

Women economists have not only critiqued established theory, methodology and policy approaches in current economics but have also pushed for and produced gender-aware questions and analyses and a more inclusive program of research. They have also sought more pluralism in methodology and research methods. While they have not yet had the transformative effect in economics that women archaeologists and medical researchers have had in their fields, women economists are becoming more widely recognized for their contributions, and their work is gaining increased recognition in places like the United Nations (see, e.g., Barker and Feiner 2004).

These are some of the facts relevant to the Larry Summers debate that can be gleaned from the critiques described above. What shall we conclude? The women's work in medical research and archaeology and economics that I have described (and similar work in other areas of science that I could have described) is not only epistemologically impressive – productive of science that is more accurate, more thorough, more comprehensive and better justified than the men's work that preceded it; it is also politically impressive – productive of science that is more egalitarian, more fair-minded and more helpful to more people than what preceded it. This work of women scientists, in short, is importantly different from that of their male counterparts, but certainly no less significant. Were these women scientists never a part of science, our science and the society moulded by our science would be in far worse shape. Women's lives, in particular, would be in far worse shape.

Of course, not all women scientists have done the kind of path-breaking research described above, and not all men scientists have not. There have been women scientists whose work was indistinguishable from that of the men, women scientists whose work was just as biased and limited as that of the men. At the same time, there have been men scientists who have done the same kinds of groundbreaking critical and constructive work done by the women, and certainly men scientists who have done other kinds of groundbreaking critical and constructive work. Still, the remarkable pattern of achievement of women researchers described above remains. And, of course, the women scientists I have described hardly achieved their breakthroughs single-handedly. The changes in science they brought about drew upon many resources: a broad-based women's movement and the fundamental changes in attitudes toward women and their place in society and the professions that this movement achieved; a related women's health movement and the legislation that mandated gender- and race-related changes in medicine; the institutionalization of academic research on gender together with the interdisciplinary collaborations that this made possible; strong lobbies on issues of public concern, such as breast cancer; and so on. But most of these resources were also provided by women. Shall we, then, conclude, in Larry Summers fashion, that some scientists *are* more gifted, more intelligent, more motivated in more valuable directions, than other scientists, but that these superior scientists are the women, not the men?

Reframing the debate, part 2: adding the insights of feminist philosophy of science

Perhaps focusing attention, as Summers does, exclusively on intelligence and motivation does not provide a rich enough conceptual base for understanding the contributions of women scientists and the significance of the gender gap in science. But what will do better? Traditional resources for dealing with issues concerning science and knowledge, such as epistemology and philosophy of science, offer little promise, for they tend to ignore gender altogether – indeed, they have historically pursued programs of research that make it difficult not to ignore gender. For example, Descartes and the sceptical challenges he posed in the *Meditations* as well as other works have done much to define the programs of research of modern epistemology, but these programs have centrally concerned attempts to answer the sceptic – attempts (on the part of a genderless 'mind') to prove the existence of 'other minds', the 'external world' (including one's own body), the past and so on – hardly a foundation for the issues the gender gap represents. And the programs of research in mainstream philosophy of science, given its goal of articulating 'scientific rationality', have concerned themselves with such topics as the logic of scientific explanation, the patterns of scientific development and the relation of scientific knowledge to the world, again hardly a foundation for dealing with the gender gap in science. By contrast, the new areas of feminist epistemology and philosophy of science, like all other feminist projects, have been concerned with women and their struggle for equality and hence have aimed as a matter of course to make gender visible. So they are far more likely to be sources of help for us here.

Consider feminist philosophy of science in particular, especially the work of Sandra Harding and Helen Longino, two of the founders of the field and certainly its most distinguished practitioners. Harding, past director of the UCLA Center for the Study of Women and currently a professor of education and women's studies at UCLA, has authored or edited more than 13 books, including the groundbreaking *The Science Question in Feminism* (1986), *Whose Science? Whose Knowledge? Thinking from Women's Lives* (1991), and *Is Science Multicultural? Postcolonialisms, Feminisms, and Epistemologies* (1998). Harding has lectured at over 200 universities around the world and served as a consultant for such international agencies as the Pan-American Health Organization, the UN Development Fund for Women (UNIFEM), the UN Educational, Scientific and Cultural Organization (UNESCO) and the UN Commission on Science and Technology for Development. She specializes in postcolonial theory, epistemology and research methodology, as well as feminist philosophy and philosophy of science. Longino, Clarence Irving Lewis Professor of Philosophy and chair of the Philosophy Department at Stanford University, specializes in philosophy of biology and social epistemology, as well as feminist philosophy and philosophy of science. She taught at Mills College, Rice University and the University of Minnesota before joining the Stanford faculty, and she helped develop women's studies programs at those institutions. She has published widely in both feminist and mainstream science studies, and her books *Science as Social Knowledge* (1990) and *The Fate of Knowledge* (2002) have been highly influential. She is currently completing a book on contemporary scientific approaches to the study of human behaviour. What insights can Longino and Harding offer regarding women scientists and the gender gap in science? (For what follows, see especially Harding 1986, 1991; and Longino 1990, 1993, 2002.)

Bypassing the qualms of Descartes and his long line of epistemological devotees, both Harding and Longino emphasize the embodied nature of scientists and their resulting 'situatedness'. According to these philosophers, that is, scientists have bodies situated at any time in particular social/cultural (gender, racial/ethnic, class, sexual-orientation, etc.) as well as physical locations. As a result, scientists conduct their research from particular – and different – spatial-temporal-social/cultural vantage points, and these can have a decided effect on the nature of that research. Though scientists might be trained in comparable ways and might use comparable research methods, neither the training nor the methods can be guaranteed to screen out the difference in vantage points from which scientists approach their research. Indeed, such vantage points, and the histories and interests and values and sensitivities they incorporate, can and do affect which questions scientists investigate and which they ignore, which background assumptions they accept and which they reject, which observational or experimental data they select to study and the way they interpret those data, and so on. It should come as no surprise, then, that the women archaeologists and medical researchers and economists described above did science differently from their male colleagues – challenged old assumptions (such as the 'natural' division of labour between the sexes), raised new questions (e.g., regarding the needs and priorities of women in families), devised new hypotheses (such as the hypothesis that women were the main providers of the staples in ancient diets) and made use of new concepts (such as the concept of heart disease as a female as well as male disease). But why should these new assumptions and questions and hypotheses and concepts have constituted *better* science rather than just *different* science? That is, why should the women scientists have made the science to which they contributed more accurate, more thorough, more comprehensive and better justified than what had been developed by the men? The answer, according to Longino and Harding, relates to the conditions of scientific objectivity.

For Longino, scientific objectivity in an important sense has to do with limiting the intrusion of any individual scientist's subjective input into the scientific community's shared beliefs – its 'knowledge' – and, hence, depends on the extent to which any scientist's particular vantage point and resulting scientific work undergoes critical scrutiny. 'It is the possibility of intersubjective criticism . . . that permits objectivity' (1990, 71); 'the objectivity of scientific inquiry is a consequence of this inquiry's being a social, and not an individual, enterprise' (67). But objectivity in this sense is a matter of degree. More specifically, it depends on the degree to which a scientific community satisfies four conditions. First, the community must have public venues for criticism, such as journals and conferences. Second, it must have publicly recognized standards – shared values as well as substantive principles – by reference to which the criticism can be made. Third, it must be responsive to the criticism. That is, the beliefs of the community as a whole and over time – as measured by such public phenomena as the content of textbooks, the distribution of grants and awards and the flexibility of dominant world views – must change in response to the critical discussion taking place within it. And fourth, the community must recognize the equal intellectual authority of all the parties qualified to engage in the debate, among whom all relevant points of view that can serve as sources of criticism must be represented. A science will be objective, then, to the degree that it satisfies these four conditions – to the degree that it permits what Longino calls 'transformative criticism'. Since the women archaeologists and medical researchers and economists described above were new additions to their fields (or old additions with newly recognized intellectual authority) and since they approached their fields from vantage points different from the men's, they provided new points of view from which to criticize the men's contributions and, hence, increased the objectivity of the resulting science. Small wonder the science was better than before.

Harding offers a different analysis. According to Harding, just as the various spatial-temporal-social/cultural vantage points of scientists can have a decided effect on what they understand and can contribute, the various vantage points of everyone else can have a decided effect on what they understand and can contribute. But these various vantage points are not always equally valuable, epistemologically. Individuals who are in the socially disadvantaged positions in society are often able to recognize more readily than those in the more advantaged positions the structures that keep in place the hierarchy of advantage and disadvantage. 'They have less to lose by distancing themselves from the social order; thus, the perspective from their lives can more easily generate fresh and critical analyses' (Harding 1991, 126). Thus, the wheelchair-bound person is painfully aware of the architectural choices and conventions (e.g., stairs and escalators rather than elevators) that disenable her mobility while they enable the mobility of the 'abled'; the abled are oblivious to all this. Gays and lesbians are aware of the heterosexual expectations and customs that deny their sexuality, while straights comfortably take them for granted; women continue to be amazed by the sexism that men fail to see; and so on. This is especially true if the wheelchair-bound, the gays and lesbians and the women have been engaged in the kinds of consciousness-raising group activities and political activism that have characterized recent movements for social equality, such as the civil rights, gay rights and disability rights movements and the women's movement. 'Only through such struggles can we begin to see beneath the appearances created by an unjust social order to the reality of how this social order is in fact constructed and maintained' (Harding 1991, 127). Such struggles help to create a more collective vantage point that Harding calls a standpoint. Thus, we can speak of a women's standpoint or a gay and lesbian standpoint or a disability standpoint. Of course, men, straight persons and abled persons can learn – have learned – to see things from these less partial, less distorted, collective vantage points, these standpoints, but the ones whose standpoints they are will typically have to be their teachers and will still tend to be the epistemological path-breakers.

So some vantage points – those associated with social disadvantage – can bring with them epistemological advantage. But this holds in science as well as out of it. Women scientists, for example, though they may enjoy the class-related and other social advantages associated with being scientists, still struggle both in and out of science with the gender-related disadvantages associated with being women. And this can provide them with vantage points on their fields less distorted than the ones available to their male colleagues. Small wonder it was the women scientists and not the men who uncovered and helped to rectify the gender-related shortcomings in archaeology, medical research and economics described previously. This, at any rate, is Harding's analysis of the situation.

What follows from the above? For both Harding and Longino, the distinctive vantage points from which women scientists pursue their research are crucial to achieving objectivity in such fields as archaeology, medical research, economics and the other fields in which gender is relevant to the subject matter of the field. But whereas for Longino this objectivity is achieved through the increased community dialogue to which women's vantage points contribute, for Harding it is achieved through the decreased distortion present in the women's vantage points themselves. Either way, the contributions of women scientists are necessary if genuine scientific knowledge is to be had. Longino explains: 'Science is thought to provide us with a view of the world that is objective in two seemingly quite different senses of that term' (Longino 1990, 62). In one sense scientific objectivity has to do with the truth of scientific claims to knowledge. In the second sense scientific objectivity has to do with the distinctive procedures scientists use to obtain that knowledge – what many call *scientific method* or *scientific rationality*. 'Common wisdom has it that if science is objective in the first sense it is because it is objective in the second' (63). So what the above shows, stated more precisely, is that women's contributions are crucial to achieving sound scientific methods in such fields as archaeology, medical research and economics, where these sound scientific methods are what produce genuine scientific knowledge in these fields. Longino also suggests, however, that the world may be so complex that a multiplicity of approaches will be required to capture all its various aspects. That is to say, a pluralism in the conduct of inquiry, the pluralism that Longino holds to be methodologically necessary, may also yield as its final outcome an irreducible pluralism of representations, a pluralism that includes women's distinctive contributions. But even if the final outcome of inquiry is a single unified representation of the world rather than a pluralism of representations, that single representation may still include women's distinctive contributions. Either outcome would furnish an additional reason women's contributions are crucial to science.

The debate reframed

None of this necessarily ends the debate over the gender gap in science. Summers can still point out that the gender gap is most prominent in precisely those areas in which gender is irrelevant – the physical and engineering sciences. So even if women's equal participation is crucial to the sciences in which gender *is* relevant – the biological and social sciences – nothing follows regarding the gender gap in general. Some scientists have argued, however, that the military aims shaping much of the research in the physical and engineering sciences are a very masculine affair and a very big threat to world peace and human flourishing (see, e.g., Easlea 1987; Cohn 1987 and 2007). Were more women to enter those fields, they have surmised, the nature of the research that goes on in them might undergo significant change (in Harding and Longino's terms, research conducted from women's vantage points might form a significant critique of the research done by the men and might even transform those fields in socially and epistemically fruitful ways). So women's equal participation might be important even to the physical and engineering sciences. Of course, Summers could complain, and rightly so, that all this is purely speculative: there is simply no way to know what effects women would have on the physical and engineering sciences were the gender gap closed. But there was also no way to know what effects women would have on fields such as archaeology and medical research before the gender gap in those fields was narrowed. This should be no surprise: we can come to know women's potential only if we do close, or at least significantly narrow, the gender gap in science.

So the debate might continue, but focused now, not on who is more intelligent than whom or who is more motivated, but on the benefits that women have brought to science and still can bring and the conditions that can help them to do just that. This would be a more fitting, more inspiring kind of debate than the one Summers started, however, and a more useful one as well – just one of the good things in store when we take advantage of the resources of feminist philosophy of science.

Notes

1 Remember that these persons are rational, autonomous, self-interested persons – what men are supposed to be by the norms of masculinity; they are not the emotional, social, other-directed persons that women are supposed to be by the norms of femininity.

2 'Tradition, in particular, may be a far more powerful force in determining the allocation of household tasks than rational optimization' (Ferber and Nelson 1993, 6).

References

Barker, Drucilla K. and Susan F. Feiner (2004), *Liberating Economics: Feminist Perspectives on Families, Work, and Globalization*. Ann Arbor: University of Michigan Press.

Baxter, Jane Eva (2005), *The Archaeology of Childhood: Children, Gender, and Material Culture*. Walnut Creek, CA: AltaMira.

Ceci, Stephen J. and Wendy M. Williams (eds) (2007), *Why Aren't More Women in Science? Top Researchers Debate the Evidence*. Washington, DC: American Psychological Association.

— (2010), *The Mathematics of Sex: How Biology and Society Conspire to Limit Talented Women and Girls*. New York: Oxford University Press.

Cohn, Carol (1987), 'Sex and Death in the Rational World of Defense Intellectuals', *Signs: Journal of Women in Culture and Society*, 12(4): 687–718.

— (2007), 'Wars, Wimps, and Women: Talking Gender and Thinking War', in Michael S. Kimmel and Michael A. Messner (eds), *Men's Lives* (7th edn.). Boston: Pearson Education.

Conkey, Margaret W. (2003), 'Has Feminism Changed Archaeology?' *Signs: Journal of Women in Culture and Society*, 28(3): 867–80.

— (2008), 'One Thing Leads to Another: Gendering Research in Archaeology', in Londa Schiebinger (ed.), *Gendered Innovations in Science and Engineering*. Stanford, CA: Stanford University Press, pp. 43–64.

Conkey, Margaret W. and Sarah H. Williams (1991), 'Original Narratives: The Political Economy of Gender in Archaeology', in Micaela di Leonardo (ed.), *Gender at the Crossroads of Knowledge: Feminist Anthropology in the Postmodern Era*. Berkeley and Los Angeles: University of California Press, pp. 102–39.

Easlea, Brian (1987), 'Patriarchy, Scientists and Nuclear Warriors', in Michael Kaufman (ed.), *Beyond Patriarchy*. Toronto: Oxford University Press.

Estin, Ann Laquer (2005), 'Can Families Be Efficient? A Feminist Appraisal', in Martha Albertson Fineman and Terence Dougherty (eds), *Feminism Confronts Homo Economicus: Gender, Law, and Society*. Ithaca, NY: Cornell University Press.

Ferber, Marianne A. and Julie A. Nelson (eds) (1993), *Beyond Economic Man: Feminist Theory and Economics*. Chicago and London: University of Chicago Press.

— (2003), *Feminist Economics Today: Beyond Economic Man*. Chicago and London: University of Chicago Press.

Gallagher, Ann M. and James C. Kaufman (eds) (2005), *Gender Differences in Mathematics: An Integrative Psychological Approach*. Cambridge: Cambridge University Press.

Gura, Trisha (1995), 'Estrogen: Key Player in Heart Disease among Women', *Science*, 269(5225) (August 11): 771–3.

Harding, Sandra (1986), *The Science Question in Feminism*. Ithaca, NY: Cornell University Press.

— (1991), *Whose Science? Whose Knowledge? Thinking from Women's Lives*. Ithaca, NY: Cornell University Press.

— (1998), *Is Science Multicultural? Postcolonialisms, Feminisms, and Epistemologies*. Bloomington and Indianapolis: Indiana University Press.

Joyce, Rosemary (2000), 'Girling the Girl and Boying the Boy: The Production of Adulthood in Ancient Mesoamerica', *World Archaeology*, 31(3): 473–83.

Longino, Helen (1990), *Science as Social Knowledge: Values and Objectivity in Scientific Inquiry*. Princeton, NJ: Princeton University Press.

— (1993), 'Subjects, Power, and Knowledge: Description and Prescription in Feminist Philosophies of Science', in Linda Alcoff and Elizabeth Potter (eds), *Feminist Epistemologies*. New York: Routledge, pp. 101–20.

— (2002), *The Fate of Knowledge*. Princeton, NJ, and Oxford: Princeton University Press.

Mann, Charles (1995), 'Women's Health Research Blossoms', *Science*, 269(5225) (August 11): 766–70.

Meinert, Curtis L. (1995), 'The Inclusion of Women in Clinical Trials', *Science*, 269(5225) (August 11): 795–6.

Meskell, Lynn (1998), 'Intimate Archaeologies: The Case of Kha and Merit', *World Archaeology*, 29(3): 363–79.

MIT Committee on Women Faculty in the School of Science (1999), 'A Study of the Status of Women Faculty in Science at MIT', special edition of the *MIT Faculty Newsletter*, 11(4), http://web.mit.edu/fnl/women/women.html.

Moss, Kary L. (ed.) (1996), *Man-Made Medicine: Women's Health, Public Policy, and Reform*. Durham, NC, and London: Duke University Press.

Nelson, Julie A. (1996a), *Feminism, Objectivity and Economics*. London and New York: Routledge.

— (1996b), 'The Masculine Mindset of Economic Analysis', *Chronicle of Higher Education*, 42(42): B3.

Rosser, Sue (1994), *Women's Health – Missing from U.S. Medicine*. Bloomington and Indianapolis: Indiana University Press.

Schiebinger, Londa (1999), *Has Feminism Changed Science?* Cambridge, MA: Harvard University Press.

Schmidt, Robert and Barbara Voss (eds) (2000), *Archaeologies of Sexuality*. London: Routledge.

Sherman, Linda Ann, Robert Temple and Ruth B. Merkatz (1995), 'Women in Clinical Trials: An FDA Perspective', *Science*, 269(5225) (August 11): 793–5.

Summers, Lawrence (2005), 'Remarks at NBER Conference on Diversifying the Science and Engineering Workforce', www.president.harvard.edu/speeches/summers_2005/nber.php.

Waring, Marilyn J. (1992), 'Economics', in Cheris Kramarae and Dale Spender (eds), *The Knowledge Explosion*. New York and London: Teachers College Press, pp. 303–9.

— (1997), *Three Masquerades: Essays on Equity, Work, and Hu(man) Rights*. Toronto: University of Toronto Press.

Watson, Patty Jo and Mary C. Kennedy (1991), 'The Development of Horticulture in the Eastern Woodlands of North America: Women's Role', in Joan M. Gero and Margaret W. Conkey (eds), *Engendering Archaeology: Women and Prehistory*. Oxford and Cambridge, MA: Basil Blackwell, pp. 255–75.

Weisman, Carol S. and Sandra D. Cassard (1994), 'Health Consequences of Exclusion or Underrepresentation of Women in Clinical Studies (I)', in Anna C. Mastroianni, Ruth Faden and Daniel Federman (eds), *Women and Health Research*, vol. 2. Washington, DC: National Academy Press, pp. 35–40.

Wilkie, Laurie (2003), *The Archaeology of Mothering: An African-American Midwife's Tale*. London: Routledge.

Wylie, Alison, Janet R. Jakobsen and Gisela Fosado (2007), *Women, Work, and the Academy: Strategies for Responding to 'Post–Civil Rights Era' Gender Discrimination*. New York: Barnard Center for Research on Women, http://feministphilosophers.files.wordpress.com/2008/01/bcrw-womenworkacademy_08.pdf (accessed February 24, 2008).

AFTERWORD

We have seen some of the key thinkers and great philosophers of science and looked at many of the main themes that concerned them. Now it is time to acknowledge the gaps and to gaze into the crystal ball to see what the future holds.

The gaps are many; some I mentioned in the opening chapter. Philosophy of science is a wonderfully rich field, filled with stellar philosophers trying to make sense of it. We have seen only a sample from a large population. A book several times bigger would still not have done full justice to the greats of the past or to the greats of today. Readers, I'm sure, will understand the constraints of space limitations. As well as having to forgo this or that particular philosopher, there are whole classes of philosophers of science who were omitted – namely, philosophers of the special sciences. Philosophers of physics, of biology, of the foundations of probability, of the social sciences and of mathematics have all been ignored. Instead of including them, this volume has been restricted to a selection of key thinkers who are general philosophers of science. To do otherwise would have diluted things far too much.

It is regrettable, since specialization is on the rise within the philosophy of science – a trend that has been in effect for some time. The philosophers of biology, philosophers of physics, philosophers of mathematics and so on, are growing rapidly in number. Typically they have attained a high level of expertise in their respective branches of science, and they often publish in the same journals as working scientists. Some of this is reminiscent of the very best work of the seventeenth century, when philosophy and the sciences were intimately intertwined to the benefit of each.

In general philosophy of science, the current range of topics will doubtless continue to develop. Scientific realism, a perennial topic, has evolved into a clash over so-called structural realism. The main idea is that it is the underlying structure proposed by a theory, not the particular details, that we should be realists about and that over time it is structure that survives in transitions, say, from classical to quantum mechanics. The nature of explanation remains a topic of great interest, with much debate between those who like a 'top-down' explanation of things, in terms of fairly abstract laws of nature, against those who prefer a 'bottom-up' account, in terms of mechanisms. Models play a role here and elsewhere. The range of models is enormous, and classifying their use is one of many current activities. Some seem like rough, tentative theories; others are concrete entities (wooden car in a wind tunnel); yet others are quite abstract (mathematical models). They function in different ways. Figuring these out is a major occupation of many current philosophers of science.

A topic I myself am particularly interested in is thought experiments. While it is widely acknowledged that they play a huge role in philosophy and the sciences, especially physics, only recently have they become the subject of investigation in their own right. How do they work? What are the different kinds? How realistic do they have to be? The big question is simply this: How is it possible that merely by thinking we seem to learn new things about the world? The answers range from the claim that we have the cognitive capacity to be able, a priori, to grasp some laws of nature to the claim that we are really just deriving conclusions from empirically justified premises. The field of thought experiments is growing fast in adherents and in sophistication.

Philosophers of science, like philosophers generally, have upheld a fact-value distinction and claimed that good science is value free. This is rapidly breaking down. Now there are numerous studies of value in science, usually claiming that this is essential to science and that it need not mean that science has lost its objectivity. Feminist philosophers of science have been very influential over recent decades, and here is one of the places where their influence is most keenly felt.

There is an uneasy relationship between philosophers of science and those variously called sociologists of science, social constructivists and so on. Philosophers, as I mentioned in the introductory chapter, tend to see science as a rational and progressive activity. Philosophers may fight over the particular details of rationality, but not the idea that some kind of objective evidence is at work. They see social constructivists as claiming the course of science is due to social or psychological factors and that scientific decision making is based on anything but rational factors.

By contrast, social constructivists see philosophers of science as naive cheerleaders for science, ignorant of the real factors that drive science along. Both camps are at least partly right, and each could profit from paying closer attention to the other. The best thing any philosopher can do is immerse herself in the history of science, then try to sort these issues out for herself.

These brief remarks are, of course, just a guess at what topics are likely to interest philosophers of science in the near future. I recommend looking at some of the main journals to get a feel for what is going on.

Biology and Philosophy
British Journal for the Philosophy of Science
International Studies in the Philosophy of Science
Philosophy of Science
Studies in the History and Philosophy of Science
Studies in the History and Philosophy of Biology
Studies in the History and Philosophy of Physics

And when it comes to particular topics, the internet continues to provide ever better sources of information. One of the best is the online *Stanford Encyclopedia of Philosophy*. There one can find excellent articles on explanation, realism, models, Carnap, philosophy of biology and so on, and each will have a very thorough and up to date bibliography.

I want to end on a somewhat personal note by responding to two questions: Is the philosophy of science in a healthy state?, and, Is it a useful thing to do? To each of these I am inclined to utter a decisive and unequivocal 'Um, well, sort of'. There is no question that philosophers of physics have contributed greatly to current physics. The so-called hole argument, for instance, tells a lot about General Relativity, and it also may play a role in the future development of quantum gravity. Philosophical interest in the problems of quantum mechanics was long dismissed by physicists as 'mere metaphysics', but these issues have become central to quantum computation and quantum cryptography. In debates over the nature of biological species, philosophers seem to dominate the discussion. In archaeology, anthropology, psychology, economics and so on, philosophical input has been of considerable importance. This is all to the good – for philosophy, for physics, for biology and for all the other sciences. It is fair to say that philosophy cannot be avoided – it can only be done well or poorly.

Where current philosophy of science might be falling down is in relation to mainstream metaphysics and epistemology. Analytic M&E, as it is often called, can be downright brilliant. But very little of it has any application to problems in the philosophy of science. Solving the problem of scepticism or saying what knowledge is in a Gettier-type situation is all well and good, but such achievements shed no light on the most impressive examples of knowledge humanity possesses: namely, the acceptance of Darwinian evolution, the passage from classical to quantum mechanics and other such cases in the history of science. If analytic epistemology cannot explain the history of science, then it is worthless. On the other hand, the sheer brilliance of some contemporary metaphysics and epistemology will surely help to explain what is going on in the sciences.

The other place where contemporary philosophy of science might fall short of the ideal involves its relations to society and social issues. No one is as well suited as a smart philosopher of science to address some of the major issues involving science and society. Who better to take on the silliness of so-called intelligent design, to evaluate the charges made by climate-warming deniers or to investigate the effects of commercialization on the quality of pharmaceutical research? Of course, working scientists should be playing a bigger role here as well, but philosophical evaluation is particularly important, since often questions involving the very nature of evidence are at issue. That's *our* turf. Playing such a role in society, as well as doing our everyday research, would not be new. Plato urged philosophers to become kings or teach kings to become philosophers. Members of the Vienna Circle, as I mentioned in the introductory chapter, were very politically active, fighting Nazis among other things. When they emigrated to the United States, however, they dropped their political activity and focused on other issues relating to science. But the need for serious involvement of philosophers and of academics generally has never been greater.

Marx famously said, 'Philosophers have only interpreted the world; the point is to change it.' Of course, we are unlikely to change it without acquiring some level of understanding first. One of the best places to begin is with an understanding of science and of the key thinkers who can unlock that door for us.

Index